NATO ASI Series

Advanced Science Institutes Series

A series presenting the results of activities sponsored by the NATO Science Committee, which aims at the dissemination of advanced scientific and technological knowledge, with a view to strengthening links between scientific communities.

The Series is published by an international board of publishers in conjunction with the NATO Scientific Affairs Division.

A	Life Sciences	Plenum Publishing Corporation
B	Physics	London and New York
C	Mathematical and Physical Sciences	Kluwer Academic Publishers Dordrecht, Boston and London
D	Behavioural and Social Sciences	
E	Applied Sciences	
F	Computer and Systems Sciences	Springer-Verlag Berlin Heidelberg New York
G	Ecological Sciences	London Paris Tokyo Hong Kong
H	Cell Biology	Barcelona Budapest
I	Global Environmental Change	

NATO-PCO DATABASE

The electronic index to the NATO ASI Series provides full bibliographical references (with keywords and/or abstracts) to more than 30000 contributions from international scientists published in all sections of the NATO ASI Series. Access to the NATO-PCO DATABASE compiled by the NATO Publication Coordination Office is possible in two ways:

- via online FILE 128 (NATO-PCO DATABASE) hosted by ESRIN, Via Galileo Galilei, I-00044 Frascati, Italy.

- via CD-ROM "NATO Science & Technology Disk" with user-friendly retrieval software in English, French and German (© WTV GmbH and DATAWARE Technologies Inc. 1992).

The CD-ROM can be ordered through any member of the Board of Publishers or through NATO-PCO, Overijse, Belgium.

The ASI Series Books Published as a Result of
Activities of the Special Programme on
ADVANCED EDUCATIONAL TECHNOLOGY

This book contains the proceedings of a NATO Advanced Research Workshop held within the activities of the NATO Special Programme on Advanced Educational Technology, running from 1988 to 1993 under the auspices of the NATO Science Committee.

The books published so far as a result of the activities of the Special Programme are as follows (further details are given at the end of this volume):

Control Technology
in Elementary Education

Edited by

Brigitte Denis

Service de Technologie de l'Education
Université de Liège au Sart-Tilman, Bât. B32
B-4000 Liège 1, Belgium

Springer-Verlag
Berlin Heidelberg New York London Paris Tokyo
Hong Kong Barcelona Budapest
Published in cooperation with NATO Scientific Affairs Division

Proceedings of the NATO Advanced Research Workshop on Control Technology in Elementary Education, held in Liège, Belgium, November 17–21, 1992

CR Subject Classification (1991): K.3.1, D.1.3, J.2

ISBN 3-540-56710-0 Springer-Verlag Berlin Heidelberg New York
ISBN 0-387-56710-0 Springer-Verlag New York Berlin Heidelberg

© Springer-Verlag Berlin Heidelberg 1993
Printed in Germany

Typesetting: Camera-ready by authors
40/3140 - 5 4 3 2 1 0 - Printed on acid-free paper

Preface

Control technology is a new learning environment which offers the opportunity to take up the economic and educational challenge of enabling people to adapt to new technologies and use them to solve problems. Giving young children (and also adults) easy access to control technology introduces them to a learning environment where they can build their knowledge across a range of topics. As they build and program their own automata and robots, they learn to solve problems, work in collaboration, and be creative. They also learn more about science, electronics, physics, computer literacy, computer assisted manufacturing, and so on.

This book, based on a NATO Advanced Research Workshop in the Special Programme on Advanced Educational Technology, presents a cross-curricular approach to learning about control technology. The recommended methodology is active learning, where the teacher's role is to stimulate the learner to build knowledge by providing him/her with appropriate materials (hardware and software) and suggestions to develop the target skills. The results are encouraging, although more tools are needed to help the learner to generalize from his/her concrete experiment in control technology as well as to evaluate its effect on the target skills. The contributions not only discuss epistemological controversies linked to such learning environments as control technology, but also report on the state of the art and new developments in the field and present some stimulating ideas.

Acknowledgements

This Advanced Research Workshop could not be held without the support of the NATO Scientific Affairs Division and the Fonds National de la Recherche Scienfique of Belgium.

We would like to acknowledge Professeur Dieudonné Leclercq (director of the Educational Technology Department at the University of Liège, Belgium) for his scientific support and for the critical re-reading of this book.

We also want to thank Ingeborg Frank, the main secretary of the ARW, for the great competence she displayed before, during and after the meeting.

And finally, thanks to to all the colleagues who participated actively in the workshop and agreed to write this book.

June 1993   Brigitte Denis

Table of Contents

3.2. Future Trends

Editor's Introduction

This part of the book introduces the concept and the framework of control technology, the challenges this topic can take up and the structure of the book itself.

Control Technology (CT) is an activity which is going to be more and more used in education. Three international conferences have been organized during the last years on this theme (Le Mans, 1989; Montréal, 1990; Mexico, 1991). The fourth one will take place in Belgium (Liège) in July 1993.

The success of this kind of topic is attributed to the fact that Control Technology deals not only with technology (robotics) but also with education and different subject contents.

1 What is Control Technology?

This book deals with Control Technology (CT) in elementary education. When the experts who participated in the ARW were asked "what is for you a CT learning environment?", they answered first that CT is not an end in itself but a contribution to learning environments in general. Some of them distinguished CT in industry from CT in education. *CT in industry* is about transforming reality and *CT in education* is about learning about transforming reality:
  – learning about the rules of reality,
  – learning from feedback from attempts to transform reality,
  – learning to program the computer in control experiments (use of a language to control an event).
In fact, everybody agrees that a CT environment can be seen as a powerful form of "intellectual mirroring".

In this learning environment children are expected to build their knowledge, to be creative in referring to their own view of the world. They play with new ideas and technologies in a concrete way since CT offers a bridge between abstraction and reality. They develop problem solving strategies, creativity, socialization. It is a microworld which permits them to explore topics such as systemic thinking and cybernetics. Different notions are discovered, used and understood by the students, related to science concepts, computer language, technology,... It is an interdisciplinary approach to education.

2 Control Technology Takes up Challenges

In fact, Control Technology could help to take up three kinds of challenges:
  – an economic one,
  – an epistemological one, and
  – a pedagogical one.

2.1 The Economic Challenge

Recent studies have shown that European and North American countries suffer from a lack of qualifications in schools and industries. For instance, IRDAC (Industrial Research and Advisory Committee of the Commission of the European Communities) has reported that to provide an increasing growth to the EEC countries, *NIT* must be taken into account because it promotes competitiveness and the demand for human resources. Provisions for SME's needs and their growth are lacking.

The relation between education and competitiveness is vital. The nature of this relation evolves because of technological mutations and the development of higher competition.

IRDAC is convinced that for the welfare of citizens, education and training linked to competency and competitiveness are of great importance.

For the future, in all industrialized countries, everything indicates that a higher demand for engineers, scientifics and technicians will be observed. There is a threat that educational systems will not be able to train sufficient numbers, with the necessary variety, of secondary and high school students.

If strategies and appropriate decisions are not made urgently, the situation will deteriorate during the first decade of the 21st century. Worse, deficits (lacks) of qualifications could be so important that this will be prejudicial for the American and European competition with other countries (e.g. Japan).

But increasing the quality and the quantity of workers and providing them a strong training is not enough. Today and tomorrow, non-technicians must be able to understand better technologies. This is particularly true for directors who are decision makers confronted by new technologies.

Another aspect is that if a more systematic exploitation of technology offers new opportunities to grow, it also requires some new considerations on resource exploitations. Technological mutations imply that companies must have competent workers. This suggests that we think about human resources and training strategies inside of small or big companies.

Today, far too few youngsters follow scientific and technical schooltracks. So the first challenge will be to make technology attractive. CT can contribute to this goal.

2.2 Epistemological Challenge

We all need, as learners, environments that help us to explore, to build new worlds and to receive substantial feedback to reflect our mental processes and favour our metacognitive habits.

We not only need what Pierre Nonnon has termed "cognitive spectacles", in French "lunette cognitive", to see the same phenomenon with our two brain hemispheres at the same time – the global, analogical and multimedia one (the right one), and the rational, analytical and semantical one (the left one) – and to favour the links between those two hemispheres.

In addition, we need a "mirror of our mind", i.e. a reflection (in optical terms) to help us to see how we think, i.e. to favor metacognition. That is what Seymour Papert had already fully understood when he created LOGO, advocating that each of us should (and could) become an epistemologist of his own knowledge, with the help of appropriate microworld environments. Control Technology is a new opportunity for such microworlds.

By using Control Technology in elementary school, girls and boys have the opportunity to develop some new competencies. Connexions between the classroom activity with real life can be made. Students are discovering and building concretely various notions such as electricity, electronics, physics, musics, art, mechanics, mathematics, ...

2.3 Pedagogical Challenge

Since the educational strategies developed in Control Technology microworlds are those most often linked to theories of constructivism and active pedagogy, pupils are building their knowledge at their own pace. They construct robots and program them. The teacher's role is to be a catalyser of their learning, a resource among others.

So technological goals and strategic ones (as development of problem solving strategies) can be reached.

Considering Taylor's classification about uses of the computer (Tool, Tutor, Tutee), Control Technology microworld can be situated in the Tutee applications.

Whereas Control Technology can be taught in the most directive way, we all share the idea that it would thus lose its potential power of instilling a creative and experimental mood in the classroom, coming from the learners themselves.

Teachers have to be trained in managing this kind of situation. Control Technology offers a marvellous field to do it, since teacher-learner interactions are not at all the same as in the classical information transmission paradigm.

3 Structure of the Book

The goal of this ARW and of this book is to make syntheses, to concretize objectives, methodology, and results of experiments, and to look at future developments of CT, i.e. to create opportunities for a dialog between researchers on one hand and between researchers and teachers on the other hand.

This book is divided into three sections related to the following themes: theoretical aspects of learning environments, reports of experiments or case studies, and tools developed for CT.

Such a division is made for structural purposes. In fact a lot of questions are linked to each other. For example, development of new pieces of software and hardware imply new applications and extend objectives to be reached in terms of contents and skills.

3.1 Theoretical Aspects of Learning Environments

Different authors deal with theories on which are based the use of learning environments such as Control Technology. Valcke presents the theoretical bases of CT referring to "developmental psychological theories and constructivistic, cognitive theories of knowledge acquisition". He also develops a critical point of view on the optimistic expectations related to learning in an interactive learning environment. Sougné deals with recent reasoning theories and modelling of human cognition, and analyzes problem solving in control technology.

Nonnon suggests an approach to physical phenomena that the student can grasp both in a symbolic and in a sensory way. That is what he calls "cognitive spectacles". This helps the learner to access to induction and to deduction.

Doyle gives a theoretical framework related to information processing and to learning. He discusses (and even disputes) the meaning usually assigned to Control Technology and disjunctions between used language and activities.

Gyftodimos and his colleagues address the role of Control Technology within a special kind of computer use: that of the exploratory formulation and expression of ideas by learners and the ongoing debugging processes necessary to a constructivist approach. For these authors, CT provides an opportunity to represent ideas and understand principles (e.g. variables).

3.2 Experiments and Case Studies

This section presents some authors' points of view on three different topics: contents, methodology, and the teacher's role.

3.2.1 Contents

Some presentations deal with contents related to science, others are more oriented to technology itself.

Bres describes the experiments of the "Ecole active de Malagnou" where children have had the opportunity to use computers in different ways. He develops the Pangée project where the students use telecommunication to share their experience with others. He envisages piloting robots located in a given part of the world from another (remote) place.

Guitert presents a project where an interdisciplinary approach is carried out. Her proposal is to set up an automatic classroom where control technology gives teachers of different disciplines (mathematics, chemistry, physics, English,...) the opportunity to work together.

Schaffer's presentation concerns training teachers to help them use control technology to be more comfortable with science and technology. Her project is also to develop telecommunication links between schools.

3.2.2 Methodology

Different approaches can be developed referring to the main objectives of the teachers or to the learning phase. In fact we can try to develop a constructivist approach at different steps.

As an illustration, during the discussion on methodology, Eyre distinguished three learning phases. The first is an experiential one. Learners are using models, manipulating and controlling them. They are experiencing their limitations (what it will do, what it might do if, ...). Examples of working at this level are using toys, PIP, Roamer, TV, ... The second phase is modification: the learners add something to the model, they copy one, ... in other words: they have a goal. That is what they do when they use LEGO® or other materials referring first to a guide. The third phase is creation. Learners imagine a model, referring to reality or not. They are the designers.

Giovannini's longitudinal case study illustrates an interdisciplinary approach where teachers worked in cooperation with researchers and science experts. The methodology was based on the active participation and cooperation of the children. Her approach is close to Eyre's view of methodology. The teachers present models and demonstrate their function. Once they are familiar with the material, children are free to create a model, referring to guides or not. The teacher's role then is to stimulate reflections on how it works, why it reacts this way, etc.

Gargarian has a theoretical reflection on learning environment design and tools for knowledge creation. He presents the design of microworlds as the only way to learn referring to a constructivist approach. He insists on the following concepts or principles, "objects-to-think-with, powerful ideas, tools for designing, and building cultural interface between natural and technological culture", that he considers the core of developing such microworlds.

Enkenberg describes the strategies used by children working with CT. References to the real world but also such tools as situational graphs, and writing of reports are used to evaluate the results of the work done by the students.

In other respects, Eyre relates his experience and the difficulties (especially the teacher's lack of knowledge and experience of mechanisms) he has encountered while using CT in schools and while training teachers in this technology.

Robinson describes the difficulty of promoting active learning, considering what has been produced to help the teachers (documents, software, ...).

The report of the workshop organised by Vivet and Leroux illustrates some stimulating directions in which to discuss educational situations.

3.2.3 The Teacher's Role

The teacher's role in a constructivist learning environment is illustrated by Vivet, whose text discusses project based teaching. Defining the attitudes to be learned, he develops the learning process from the negotiation of a contract (what is going to be learned) to the evaluation of the results. This approach is developed in an adult training situation dealing with computer integrated manufacturing (CIM).

With a learner's personal project centered approach, Denis considers interactions between teachers and learners referring to a list of target objectives. Using some tools to observe these interactions, she evaluates whether there is a gap between the theoretical choices of the teachers and their practice. Giving the opportunity to be more realistic about what happens during the activity, she also gives the teachers the opportunity to regulate their behavior in order to be more efficient.

3.3 Tools Developed for Control Technology

When the question "are available tools (software, hardware, ...) relevant ?" was asked, participants said that one must be careful in providing the answer since a lot of tools exist. Even if they consider that in fact these tools are made relevant by their users, they also assume that these tools should be examined and compared taking into account such different parameters as the richness of levels of representation concurrently available (modes of representation and interaction).

Tools developed for CT evolve continuously. Some of them were presented during the ARW. These presentations have been classified hereafter considering on the one hand "the state of the art" and on the other hand "future trends".

3.3.1 State of the Art

Doyle reports on the activities of the workshops animated during the meeting, referring to the theme "evolution of software and hardware".

Both Doyle and Louttit present a concept keyboard which permits the user to send commands without using the traditional keyboard.

Different robots such as PIP or Roamer are presented. Their creators, respectively Louttit and Meredith point out their technical characteristics and their educational uses.

Calabrese's presentation concerns a piece of software, MARTA, which is a screen "robot" that the user can guide in a labyrinth. The microworld where MARTA travels is full of events.

In other respects, Moeller's presentation and workshop illustrates the new hardware developed by the LEGO ® DACTA company: a new interface and software as well as the didactic materials associated with the new products.

3.3.2. Future Trends

Two authors have developed systems that can help the teacher or/and the learner working with CT.

Leroux's work deals with the conception of ROBOTEACH, a prototype that allows the creation of personalized learning sessions by the teacher.

Nonnon presents the construction and design of an efficient measuring instrument, in fact a software tool which allows the student to acquire data and which allows the teacher to focus more on pedagogical objectives.

Three presentations deal with parallelism. Argles develops the concept of concurrent and parallel programming. The sofware he presents is linked to the purpose of making concurrent programming available to young children.

La Palme also takes up this idea and claims that in the learning mode, a student can, at the same time, control and program a robotic device and then explore parallel programming and control.

Darche presents a heterogeneous network of actors to help in the learning of parallelism, communication and synchronisation. In the interactive microworld he is developing there is a network of real hardware and software actors. At the end of his presentation, Darche illustrates some educational situations involving parallelism.

In summary, this book offers a broad overview of different facets of Control Technology. As mentioned above, all the lectures presented here have interrelations. Their complementary approaches should help the reader to access the state of the art and to concretize future trends during his/her work in the domain.

Conclusions

During this Advanced Research Workshop on Control Technology (CT) in elementary education held in Liège (Belgium), about thirty experts in the domain have had the opportunity to share their experience and to debate on future trends in the domain. Relevant contributions on the theoretical framework, results of experiments on learner's knowledge acquisition (scientific concepts), methodology, ... as well as presentations of new hardware and software developments gave to this meeting and to this book a good understanding of the richness of such a learning environment.

Developments and the results of experiments in control technology (CT) in education presented during this meeting have been examined. As a conclusion, the following questions will be adressed:

– What is the value of CT in elementary education?
– How can children developing knowledge with CT be assessed?
– Which kind of experiences are appropriate for elementary education?

What is the value of CT in elementary education?

CT is a concrete tool to think with and act on reality. It is a learning environment, a microworld which provides learners with concrete things to explore and helps them to develop abstraction process and different scientific and technological concepts (Gargarian, Giovannini,...). Nevertheless, Nonnon insists on the fact that the connection with science is not guaranted only by the CT activity. For this author, CT in itself cannot develop a formal structure of knowledge (e.g. to construct a schema of variables control) because there exists a barrier to passing from empirical to scientific reasoning. He assumes that this obstacle can be cleared by giving the student cognitive tools which permit him/her to have a representation of the knowledge. Such kinds of tools should be developed in relation to different concepts to help the learners to build scientific knowledge.

In addition to that, CT is a good platform to encourage active teaching, an activity which devolves the responsibility for learning from the teacher to the pupils.

How can children developing knowledge with CT be assessed?

First, we should not forget to implicate the learner him/herself in the definition of the activity objectives.

Second, if CT has currently a great success in the educational world, no systematical evaluation of the experiments developed is made. We should then be attentive to Valcke's reflections on the optimistic expectations concerning active learning and not consider just enthusiastic purposes as positive results. In fact, there is a consensus among the researchers and the practitioners that to provide a learning environment such as CT does not automatically imply effective and active teaching and learning.

All the experts who participated in the ARW agree that CT needs constant evaluation in order to lead to subsequent good practice. So how can that be achieved?

In the experiments developed here, different assessment methods are used. They depend on the teacher's or researcher's objectives. We can assess knowledge developed during CT activities by the pupils like any other knowledge (e.g. test on different notions in physics (Giovannini)). We also can observe the learners' behaviors to try to detect whether they reflect the attainment of different target objectives: social ones (e.g. do children cooperate to realize a project? (Denis)) or cognitive ones referring more to problem solving than to the mastery of contents (e.g. do the learners structure their control procedures (Sougné, Enkenberg) or do they verbalize links between their action and the reaction of the robot? (Denis)).

Assessment should be done in relation to different parameters: objectives, methodology, target population,... and without forgetting to consider what the learners have actually done during the CT activity period (Denis). That is an important condition to be able to answer questions such as "are results far from the pedagogical and technological goals?" Lot of tools and methods (grids, video, tests, ...) allow teachers to have a feedback on their practice. But other various tools are still to be developed to assess the various facets mentioned above and also to integrate links between them.

Once the attainment of objectives is assessed, there is an other crucial question: is the learning achieved during CT activity transferable to other situations?

What kind of experiences are appropriate for elementary education?

A lot of experiments and case studies showed that experiences where learners propose what they want to do, and plan and modify their action together with the teacher, are appropriate for elementary education, and also for adult training (Eyre, Vivet, Schaffer). Cooperative learning, cognitive conflict, team learning, and interdisciplinarity are important and basic concepts for working in a CT learning environment (Guitert, Giovannini, ...).

An interesting link between research and practice could be found by building pedagogical models of teaching and learning strategies. To do this, we should proceed as in scientific research: observe what is happening to understand it, develop experiments to explain it, and develop models to be able to predict some phenomena. Consequently, a strong framework could be available for researchers and practitioners.

In summary, this meeting provides relevant future trends not only in a technocentric way (the field of software and hardware developments) but also in an educational one, for practitioners and also for researchers, ensuring that the learning environment which is Control Technology has a substantial future.

Brigitte Denis
Director

1. Theoretical Aspects of Learning Environments

Knowledge Representation and the Learning Process: Taking Account of Developmental Features and Support Features in Interactive Learning Environments

M.M.A. Valcke, Open University, Centre for Educational Technology and Innovation, PO Box 2960, 6401 DL Heerlen, The Netherlands

Abstract. Extending the learning and teaching environment with computers, multimedia, control technology, ... enhances the possibilities to incite multiple and varying knowledge representations for knowledge categories from learners. Especially the development of interactive learning environments (e.g. control technology) looks appealing since, in this way, we can help learners to develop actively a large and rich variety of knowledge representations. This can help learners to get access to complex knowledge categories. At the theoretical level, this enrichment can be founded by referring to developmental psychological theories and recent constructivistic, cognitive theories of knowledge acquisition. This article elaborates this theoretical base. But this elaboration also puts forward a set of conditions in the educational setting. The latter is especially true if we pursue a gradual integration of multiple knowledge representations towards more comprehensive schemes or structures. This introduces a section in this article on doubts about the optimistic expectations that active learning and constructing personal knowledge representations is a straightforward result of learning in interactive learning environments. Guidelines are therefore put forward which educators have to take into account, and also the quality of the environment offered is discussed in this context. This critical section is followed by the elaboration of two examples that illustrate better practice. We finish the article by elaborating the remarkable parallelism between the discussion about interactive learning environments and science education.

Keywords. Interactive learning environments, knowledge representations, control technology

1 Introduction

"Interactive learning environments" represent a growing area of research, theory construction and the exchange of explorative experiences, involving a very varied audience: pupils, students, adults and teachers.

"Interactive learning environments" are in the context of this article defined as environments in which the learner is to a major extent in charge of, in control of or the main source in giving direction to his/her (or their) learning process. Interactive learning environments imply that part of the control of the learning process is passed to the learner. Technology (computers, devices, resources,...) play an important role in interactive learning environments.

A very imaginative subset of the larger family of interactive learning environments are represented by control technology applications. They represent a special sub-family since they extend the learning environment, based on a computer configuration, by adding construction kits or robot kits, comprising moving, sensoring, light-sensitive, ... elements. But also in this specific type of interactive learning environment the major feature remains: the learner is in control of major aspects of the learning process.

The particularity of interactive learning environments can also be shown when looking at the 'materials' offered to the learner to think with. In contrast to more traditional learning situations where pupils are mainly offered text-based materials, pupils get here concrete materials, construction materials, manipulative devices, etc.

    The relationship between control technology and LOGO learning environments (microworlds, programming languages) is apparent and historically the former can be considered as a further extension of the latter. At the theoretical level the same is true. This article focuses especially on this 'theoretical' issue. We probe at expliciting the explicit or implicit models of learning that are considered to represent the base for the educational practice built around educational control technology applications. Therefore learning theories that are helpful to found this type of educational practice are intensively discussed. Next, the central issue of 'knowledge representation' is dealt with. This introduces a section on doubts about the optimistic expectations that active learning and constructing personal knowledge representations are a straightforward result of learning in interactive learning environments. Guidelines are therefore put forward which educators have to take into account. The quality of the environment offered is also discussed in this context. This critical section is followed by the elaboration of two examples that illustrate better practice. The text ends on elaborating the remarkable parallelism between the discussion about interactive learning environments and science education.

2 The Learning Potential of Control Technology

In educational literature a variety of learning goals is put forward when setting up CT activities. When we analyse some typical control technology set-ups, the following learning potential can be defined.
Browning (1991) explicitly states :
- to change the attitudes of students towards mathematics & science;
- to foster cooperation among children;
- to develop mathematical concepts (estimation, measurement, proportional reasoning, graphing, the arithmetic mean, ...;
- to develop science concepts related to gears, simple machines, friction, and force.

Since the environment of Browning includes controlling devices by making use of a programming language, also learning and applying this language can be considered as a goal.

Argles (1991) defines a possible learning potential as a question :
- "What sort of learning do I want to see taking place? Am I really only about controlling things, or do I wish to develop my language and ability to think logically ?"

Rousseau, Dewez and Rorive (1991) focus on other topics :
- the idea of programming;
- the importance of splitting up problems in sub-procedures;
- the relations between man, computer, robot and environment.

Adamson and Helgoe (1989) stress completely different objectives :
- to activate artistic and creative experimentation.

We can add to this list some specific objectives related to motivational and affective attitudes.

Observing this large and varied set of objectives, a kind of déjà-vu effect emerges. Comparable lists of objectives could be compiled when reviewing the early LOGO programming literature.

In the context of this article we especially focus on the learning potential in relation to the development of conceptual knowledge. Our comments on the quality and nature of the learning incited in interactive learning settings have therefore to be positioned against these objectives: is the type of learning facilitative to reach this potential ? We do not tackle issues about problem solving strategies or the acquisition of scientific enquiry.

3 Founding Educational Practice in Learning Theory

When reading control technology research literature, advocates ground their specific approach to the learning process on a basic set of theoretical frameworks. By detecting this theoretical base we did not look for 'instructional theories'. We rather tried to put together a set of 'learning theories' that help to clarify or special focus on interactive learning environments: the development of knowledge representations. In literature explicit references to this body of theoretical literature is regularly found, next to references to instructional theories such as Bruner, Kilpatrick, Dewey, Ausubel, humanistic instructional theories, etc. But it is known that these theories build in many ways on the conceptions of the learning theories discussed.

In the further part of this text we give a short outline of the main concepts and mechanisms that help to describe each theory and finish the description by applying the concepts to learning in an interactive learning environment, e.g. control technology situations. At the end of this text part we focus on the pedagogical model that follows from these different theoretical perspectives.

3.1 Piaget

Piagetian theory is clearly developmental in nature. Successively more complex interactions with the world are the base for developing interpretive schemes. In this way the learner apprehends reality. In this interaction with the world the learner is active because he/she executes operations. The type of operations evolves depending on the age and experience level :

- sensori-motor stage	:	concrete operations (e.g. physical manipulation)
- pre-operational stage	:	first mental representations (preconceptual - intuitive thinking)
- concrete operational stage	:	logical reasoning in familiar situations
- formal operational stage	:	hypothetical reasoning

The 'schemes' are the building blocks of knowledge. Schemes are continuously used when interacting with the world. New experiences are 'assimilated' into the pre-existing scheme. If this does not work, 'accommodation' is necessary to adapt the scheme to an unpredicted/unexpected new situation. In this way 'equilibration' is reached when the new scheme fits the earlier and new realities. This gradual development of the schemes can also be described with the concept 'reflective abstraction' is the abstraction from coordinated actions. These could be elementary sensory-motor actions or more sophisticated actions such as representing the process of covering topological space (Cobb, 1987).

Slavin (1991) gives as an example to illustrate these developmental mechanisms the 'banging' of a young child on objects. Banging in the initial scheme produces a loud sound. Suddenly a baby bangs on an egg; there is no sound ... there is a silent "thud" and "splashing" ... This can have an effect on the banging-scheme. In the future the baby may bang some objects hard and others softly.

The Piagetian theory sounds readily transferrable to describe learning in an interactive learning environment : concrete manipulation with (models of) real life objects and situations are the basis for learning. Active learning is promoted: the learner has to manipulate objects, he/she is to be confronted with maladjusted personal schemes, ...

3.2 Constructivism

The structuralism of the Piagetian theory also represents the roots of constructivist theories of learning. Constructivism, as a school of thought, adheres to the following three premises (Valcke, 1990):

1. The learner actively constructs knowledge based on his/her personal experiences. Forman & Pufall (1988) operationalise this first premise by referring to three central processes: epistemologic conflict, self reflection and self regulation. The analogy with the Piagetian equilibrium model is evident. The same authors also refer to the social world outside the learner when describing these processes: "Epistemic conflict involves two knowing systems."

2. The second premise is an epistemological elaboration of the first one: If the learner constructs knowledge, does there exist 'objective knowledge'? Kilpatrick (1987) states this as follows: "Constructivism cuts the Gordian knot by separating epistemology from ontology and arguing that a theory of knowledge

should deal with the *fit* of knowledge to experience, not the match between knowledge and reality.".

3. By constructing - on the base of personal experiences - knowledge, the learner develops knowledge structures. There is an analogy with the 'scheme' concept. Important is the fact that these structures are of a dynamic nature. Vergnaud (1987) exemplifies this by referring to the concept/procedure of counting: "Counting a set is a scheme, a functional and organized sequence of rule-governed actions, a dynamic totality whose efficiency requires both sensori-motor skills and cognitive competencies.".

The constructivist view on learning is again readily transferrable to learning in interactive learning environments. Actions on objects, reality are necessary to construct the structures.

Spencer (1988) refers e.g. to "Physical knowledge" as a result of touching, beating, throwing, smelling, etc. objects. ... At the same time the experiences give the necessary experiential base to go beyond the immediacy of physical characteristics of objects. Spencer calls this "Logico-mathematical knowledge". These knowledge structures are derived from the actions : higher, faster, lower, ...

Applying this theory to learning in interactive learning environments is again a straightforward matter. Learners working with control technology get the possibility to experience concepts, mechanisms, procedures, etc. at various representation levels: they can build, manipulate, construct and reconstruct, talk about, they make plans, they explain to each other, etc.

3.3 Social Constructivism

Piagetian and constructivist theories have in literature regularly been criticised for ignoring the socially and historically situated nature of knowledge. According to these critical voices these theories deny the collaborative and social nature of meaning making (cf. O'Loughlin, 1992).

Vygotsky has a special position within the constructivistic school of thought, because in his theory the social environment gets a central place in the knowledge acquisition process. His cultural-historical theory is based on anthropological ideas.

Two basic ideas of his theory are in this context of prime importance (Cf. Wertsch, 1979):

1. Education is only successful if it relates learning to the 'zone of approximate development' (Vygotsky, 1962).
2. Vygotsky states that knowledge is constructed in an inter-individual context before internalisation can take place and knowledge becomes intra-individual : "It is claimed that cognitive strategies may first be encountered within an inter-individual context, during various joint activities. A process of internalization is proposed as the mechanism whereby these experiences become intra-individual events (Vygotsky, 1962).

A 'mechanism' playing an important role in the transition process from inter-individual knowledge to intra-individual knowledge is the "speech-act". 'Speech' is to be understood as 'language use in a social context'; as an "activity oriented conception" of language.

Applying the central ideas of the social constructivist theories to interactive learning environments we perceive that the social setting, group work is a common didactical principle used in setting up control technology settings.

3.4 Information Processing Theories

The schema-theory of Rumelhart & Norman (1981) is a typical example of an information processing theory. According to this theory the learner builds 'schemas'. Learning is an active meaning-making process where new knowledge is interpreted in terms of preexisting schemata. The nature of the schemata is very basic; it can be kinaesthetic, proprioceptive and visual.

Davies & Hersh (1981) give in this context a typical example of how giving a learner the possibility to 'manipulate' can be the clue to understand e.g. Euler's theory, concerning the relationship between the number of faces, edges and vertices of polyhedra. In their example the learner was provided with an interactive graphics system on a computer where he/she would see, rotate, change polyhedra. The visual representation was at schema level integrated with the kinaesthetic experiences to bring the hypercube to the level of intuitive understanding.

Learning is the process of schema induction where the learner generalizes from experiences particular representations of the events. Learning evolves by incorporating more and more experiences. New schema or therefore based on earlier available schemas.

The attention paid to the nature of the schemata (kinaesthetic, proprioceptive, visual, etc.) gives the clue to apply this theory to interactive learning environments. Control technology promotes active-meaning making and promotes the gradual construction of representations of differing quality and nature.

3.5 Additional Theories

The list of theoretical frameworks to base the educational practice put forward in interactive learning environments is not complete if we do not add theories that stress features such as emotions, values and aesthetics (EVA's), personal experiences, metaphors, interpretive frameworks in building representations (e.g. Bloom, 1992). Also Winn (1991) stresses this aspect of knowledge development by pointing at contextual variables and the even illogical and unpredictable nature of individual cognitive functioning.

This group of theories stresses the fact that building the 'schemes' or 'structures' is not a mere matter of semantic knowledge construction (constructing representations by inferring, elaborating, recalling, perceiving, etc.). Bloom (1992) gives an example of the latter when he refers to the disgust of some learners when learning about earthworms; anthropomorphism or zoomorphism play a part in this perspective.

This theoretical point of view fits in educational practice linked to interactive learning environments since personal involvement, playful activities, etc. are decisive features of the learning context that make the activity appealing for the learner.

In the further part of this text we will not neglect this last theoretical point of view, but nevertheless we will only elaborate on the first four frameworks.

3.6 The Pedagogical Model

If we look at the pedagogical model that follows from the different but related theoretical perspectives, then we obtain a curriculum "that is integrated so that learning occurs primarily through projects, learning centres and playful activities that reflect current interest (...)" (O'Loughlin, 1992, p.805).

No pedagogue will question the sound nature of this point of view. But the prime question emerging here is whether the learning that is incited fits the objectives pursued of the learning situation induced in the instructional setting.

Does not a difference exist between the personal 'structures', 'schemes' developed by the unique personal experiences in the interactive learning setting and the 'schemes' or 'structures' to be attained according to the predefined set of educational goals ? The nature of the representations developed in the ILE-setting is critical in this perspective.

4 A Key Concept: Knowledge Representation

A key concept in the theoretical base presented is "knowledge representation".

Advocates of interactive learning environments put forward that control technology environments can guarantee that children construct such internal representations whereby especially their nature and quality is very varied. Making this more concrete, learners in such interactive learning environments can construct the following single types of representations or combinations of them:
- motor representations (fine-motor or general)
- tactile (proprioceptive) representations
- auditive representations
- visual (3-dimensional)
- pictorial (2-dimensional)
- schematic representations
- verbal representations (descriptive, story)
- personal conceptual representations (personal 'names' for the object of phenomena)
- abstract conceptual representations (the right concept and definition)
- ...

Of course, teachers, educators will promote and pursue the development of higher level representations, such as conceptual ones. But the contribution of the theories outlined in part 3 is that they pointed out the significance and relevance of the 'lower-level' types of representations. They are the pre-conditions to provide a sound base for the so called 'higher-level' representations.

5 The Problem: From Learning to Teaching

Having confirmed the just and theoretically sound expectation that CT can be helpful to construct basic knowledge representation structures, the next question is how learning in the setting of an interactive learning environment is to be supported. If we follow the theories, knowledge representations 'will develop' to more abstract representation levels if a sufficient by rich and active environment is provided. The expectation that learners will make this transition to this more schematic, abstract level can be questioned.

When reading reports and research on implementations of interactive learning environments in a variety of educational settings, we do not get a decisive and positive answer to this question as can be seen in the following examples:

- In many control technology learning situations, learners develop a 'scheme' or 'structure' of a certain representational quality (e.g. sensori-motor experiences with differences in speed between two types of gears) but do not evolve beyond this level.
- The rich experiential base for knowledge development is also not always recognised in the subsequent learning-teaching phases. O'Loughlin (1992, p.806) describes for instance a series of research projects where "despite the apparently collaborative and active nature of the learning activities that were occurring, teachers played a very active role in defining and controlling the kinds of discourse that were permissible during the lessons.".
- The knowledge representations constructed on the base of active learning experiences can be totally inadequate when the learner is put in a new, even only slightly different, context. The transitional value (transfer-value) of the representations built so far can be limited to the initial local context. From this viewpoint, also the concept mis-conceptions can be put forward.
- Tuckey (1992, p. 273) reports positively on the learning benefits in interactive science centres. But one of her conclusions questions the overall effect of the interactive setting: "Prior knowledge was shown to be important in enabling (...) to construct plausible hypotheses about exhibits.". Prior knowledge is different between learners. Since knowledge develops by consecutively building on already existing 'schemes', 'structures', the learning outcome can be extremely different among learners. And to what extent is such learning situation manageable for educators ?
- When dealing with control technology problems, some learners cannot deal with the complexity and multi-facet nature of the entire problem. This kind of problems emerges e.g. when learners have to focus at the same time on constructing basic representations (e.g. building a motor driven car), looking at variables influencing speed, controlling the car motor, developing a control procedure in a programming environment to control the car, mastering the control programming language, filling in planning sheets, etc.
- ...

This limited but varied set of examples puts forward the need for more conscious and explicit considerations about the nature of the learning process invoked in interactive learning environments. When involving learners in interactive learning environments we expect educators:

- to be aware of the differences in prior experiences among pupils in terms of already pre-existing representations of the body of knowledge to be dealt with;
- to recognise the differences in nature and quality of representations that can result from the learning activities;
- to invoke 'conflicts' in the learning situation urging the learner to change, adapt, earlier representations;
- to go beyond the physical, sensori-motor, ... representation - regardless of their importance - by inducing verbalizations, schematic representations, formal notions and abstract language conceptions;
- to try to integrate the multiple representations of the knowledge into explicit, rich and consistent schemes;
- to plan carefully the gradual increase in complexity of the problem context put forward in the interactive learning environment;
- ...

Next to this, also the interactive learning environment itself can be an inhibiting or facilitative factor regarding this issue. If we compare e.g. the potential of a limited environment (e.g. LEGO® bricks, a motor, some bulbs and batteries) and a rich environment (LEGO® bricks, motors, bulbs, batteries, activity cards, a programming language, a drawing interface for the programming environment, ...) then we can expect that the variety of and quality of knowledge representations that can be developed from active involvement in these environments is automatically different. The problem is particularly important when a programming environment is put forward. The distance between representational possibilities of concrete LEGO-objects and the definition of this object in a programming environment can be overwhelming and a problem for learners in initial phases. Argles (1991) asks therefore for rich computer interfaces where graphical, pictorial, iconic, concepts (names), procedures, ... can be gradually constructed and interrelated.

6 Some Examples

By describing in short two examples we attempt to illustrate the two focuses of concern addressed above: the level of the educator and the level of the environment offered.

6.1 Hyper LEGO-LOGO

A collaboration between the State University Gent and the Free University of Brussels in Belgium resulted in the development of a new interactive learning environment consisting of :
- LEGO-Technik®;
- a Macintosh®-computer;
- a interface between computer and the motors, lamps and sensors of the LEGO-Technik® kit;
- a newly designed object-oriented graphical programming language.

The first three components are well known, but the fourth part of the environment is different.

The idea is that learners build, construct, compose objects with the LEGO®-bricks and the extra components of the Technik®-part. Next the learner can 'control' this object by adding wires, batteries and a switch box to the object. In the computer-environment, the learner can make a picture of these newly made objects. He can consecutively give a name to this object; e.g. he depicts a motor block and calls this 'motor' or he depicts a traffic light and calls this 'traflight'. At the same time learners can make these objects active by telling (writing down in a window) what this object can do. Here a library of 'verbs' in a reference list is helpful to document this activation of the objects. The 'controlling' of the object can be 'told' to (or 'learned' by) the computer by adding to the description of what the objects can do the way they are linked (by wiring) to the switch box.

The nice thing about the environment is that for each activity with concrete objects (bricks, motors, wires, switch box, ...) there is an alternative representation offered in the computer environment that stays close to the representations already developed by constructing instantiations of the object (e.g. a car) with the construction kit of the control of the object.

The developers of the system also impose a special structure on the learning-teaching process. In the setting learners work in teams consisting of two or three members. First, pupils agree on what they want to make. Next they start building the object (a moving bridge, a drilling machine, a mixer, ...). The task of the teacher in this phase is guaranteeing that each member has an input at this stage. What can be perceived is that in the process of negotiating and constructing, pupils try to make clear their representations to the other pupils, they 'talk', make 'pictures', use their hands to illustrate positions, movements or effects, build mini-models to make the other see the point, etc. An object is considered as finished when it can be controlled by hand or with the switch box (which the learners prefer). Depending on the experience of the pupils the process is to be repeated with the computer. They have to design, to depict the object on the screen, to "teach" the objects to the computer, to teach the computer how to control the object (or sub-objects), to design procedures to execute control sequences, etc. This phase results in the object being controlled by making use of the computer. In the next phase the team is obliged to make a large picture of the object that has been designed, possibley with additional pictures to document subparts which are not very visible from the outside. Once the picture is ready, the team has to write down a description of the object and a story of the construction process. Picture and written descriptions are passed over to other teams to check whether other understand what it is all about. This always results in discussions where one team has to clarify topics to the other teams.

This first example clearly shows how the nature of the interactive learning environment and the structure of the learning-teaching process clearly exploits the continuous construction, adaptation, reconstructions of multiple representations of concepts and processes.

6.2 LEGO®TC LOGO Unit Project

Browning (1991, p. 176) engaged a group of graduate students to design a learning unit about mathematics, making use of the LEGO®TC LOGO materials and meant for use with 4th grade children. The students wanted to develop a unit that would incorporate both science and mathematics concepts and foster a cooperative learning environment. In the following extract from Brownings' report, we especially focus on the shifts in knowledge representations the students demanded from the children. We do not outline a very systematic report.

With a LEGO®-car idea in mind (taken from activity cards), the students decided to encourage the children to think of themselves as engineers. The children would be given problems and asked of ways they believed the problems could be solved or better understood using available materials. In Lesson 1, the children were shown a LEGO car already built and provided a demonstration of the car rolling down a ramp. The problem set was how to build a car that would travel the farest down the ramp. In addition, how would they determine whose car went the farthest. After solving the measurement problem, the children started building their cars. There were not enough wheels for each group. Some groups had to use only three wheels.

After building the cars and executing mathematical measurement activities with them, the students asked how they could make their cars to run faster from a ramp. This incited adaptations to the model-cars after phases of debate, negotiating and especially experimenting: for instance making the car more heavy seemed to influence speed. .... After adding motors to the cars, the speed problem was again put as a central topic when gears were introduced.

Some important facets can be pointed out in this example. First, the students provided the children with an external representation of a 'car', a pre-built model of a LEGO®-car before asking them to build their own.

Initial representations of a 'car' were immediately questioned for instance by providing only three wheels instead of the traditional four ones. Representations about the concept of 'speed' were questioned by asking to look for variables that can be manipulated to influence the speed of the models (weight, gears, ...).

Of particular value in this example was the very careful prior planning of the construction and the excitation of shifts and changes in knowledge representations of objects and phenomena.

7 Parallelism with Science Education

Control technology as an example of interactive learning environments builds - as stated earlier - largely on earlier experiences, theories, research, ... about LOGO-like programming environments. In this tradition the 'turtle' was presented as physical object to think with, as the external emanation of the internal thinking process.

But next to this, a tradition already existed in most educational systems that also valued active learning and that also built - to a certain extent - the learning-teaching process upon the personal involvement of the pupils: science education. When reading the research body in relation to science education this parallelism is striking. Piagetian learning, constructivism, social constructivism, etc. also appear regularly in research reports, descriptions of learning situations and so on (e.g. Slone & Bokhurst, 1992; Tuckey, 1992; etc.).

The knowledge base about electricity, fluids, lenses, chemistry, ... can to a large extent be passed to pupils during practical learning sessions. Typical in the science education literature is the attention paid to the structuring of the 'active' learning situation where for instance the active learning principles as put forward earlier in this text are carefully combined with teacher interventions.

But also in this body of literature comparable problems to what was indicated earlier are encountered when developing scientific knowledge in active learning environments. Nevertheless, the long tradition in science education has been helpful to establish already a vast amount of theoretical and practical experience, sustained by a lot of empirical research. A typical example is the awareness in science education that misconceptions in knowledge representations can emerge from experiences in interactive learning environments; for example:
- motion and force (Galili & Bar, 1992)
- electricity (Saxena, 1992)
- water solution (Slone & Bokhurst, 1992)
- ...

It is therefore a pity that there is as yet still no communication between both educational traditions.

8 Conclusions

Reviewing the general rationale presented in this article we can conclude that involving learners in rich interactive learning environments places large demands on the educator planning and directing the processes and the developers of the environments.

At the same time we can indicate that an important research programme can be advocated to validate implicit assumptions about the potentiality of interactive learning environments. Topics of prime importance are:
- What types of misconceptions can emerge and how can we cope with misconceptions?
- How can we develop and validate a method to integrate multiple representations?
- How can we structure the learning-teaching situation: validation of scenarios?
- ...

We can add to the latter that bridges could be established between the educational and research tradition in science education and the new tradition building on interactive learning environments. Lessons can be learned in order to avoid recapitulating unnecessary unfruitful directions in practice and research.

References

Adamson, E. & Helgoe, C. (1989) Exploring art and technology. In: G. Schuyten & M. Valcke (eds.) Teaching and learning in Logo-based Environments. Proceedings, Gent 1989. 58-69. Amsterdam-Washington, Tokyo: IOS

Argles, D. (1991) Logo and control: two years on. In: E. Calabrese (ed.) Third European Logo Conference. Proceedings, Parma 1991. 259-286. Parma: ASI

Bloom, J.W. (1992) The development of scientific knowledge in elementary school children: a context of meaning perspective. Science Education 74(4), 399-413.

Browning, C.A. (1991) Reflections on using Lego®TC Logo in an elementary classroom. In: E. Calabrese (ed.) Third European Logo Conference. Proceedings, Parma 1991. 173-186. Parma: ASI

Cobb, P. (1987) Information-processing psychology and mathematics education - a constructivist perspective. Journal of Mathematical Behaviour 6, 3-40

Davis, P.J. & Hersh, R. (1981) The mathematical experience. Boston: Houghton Mifflin

Forman, G. & Pufall, P. (1988) Constructivism in the computer age. Hillsdale NJ: Lawrence Erlbaum

Galili, I. & Bar, V. (1992) Motion implies force: where to expect vestiges of the misconception? International Journal of Science Education 14(1), 63-81

Kilpatrick, J. (1987) What constructivism might be in mathematics education. In: J.C. Bergeron, N. Herscovics & C. Kieran (eds.). Proceedings of P.M.E. XI. 3-27. Montreal: P.M.E.

O'Loughlin, M. (1992) Rethinking science education: beyond Piagetian constructivism toward a sociocultural model of teaching and learning. Journal of Research in Science Education 29(8), 791-820

Rousseau, M., Dewez, D. & Rorive, D. (1991) How children from ages 3 to 14 progress in the classroom with Logo. In: E. Calabrese (ed.) Third European Logo Conference. Proceedings, Parma 1991. 121-128. Parma: ASI

Rumelhart, D.E. & Norman, D.A. (1981) Analogical processes in learning. In: J.R. Anderson (ed.) Cognitive skills and their acquisition. Hillsdale, NJ: Lawrence Erlbaum

Saxena, A.B. (1992) An attempt to remove misconceptions related to electricity. International Journal of Science Education 14(2), 157-162

Slavin, R.E. (1991) Educational psychology - Theory into practice. Englewood Cliffs, NJ: Prentice-Hall

Slone, M. & Bokhurst, F.D. (1992) Children's understanding of sugar water solutions. International Journal of Science Education 14(2), 221-235

Spencer, K. (1988) The psychology of educational technology and instructional media. London: Routledge

Tuckey, C. (1992) Children's informal learning at an interactive science centre. International Journal of Science Education 14(3), 273-278

Valcke, M.M.A. (1990) Effecten van het werken met Logo-microwerelden. Unpublished dissertation. Gent: State university Gent

Vergnaud, G. (1987) About constructivism. In: J.C. Bergeron, N. Herscovics & C. Kieran (eds.). Proceedings of P.M.E. XI. 42-5. Montreal: P.M.E.

Vygotsky, L. (1962) Thought and language. London: John Wiley & Sons

Wertsch, J.V. (1979) From social interaction to higher psychological processes - a clarification and application of Vygotsky's theory. Human Development 22, 1-22

Winn, P.H. (1990) Some implications of cognitive theory for instructional design. Instructional Science 19, 53-69

Reasoning Involved in Control Technology

Jacques Sougné

Université de Liège, Service de Technologie de l'Education (STE), Bât. B32, B-4000 Liège, Belgium, e-mail: Sougne@vm1.ulg.ac.be

Abstract. Problem solving in control technology is analysed in the light of the most recent reasoning theories. Divergences exist among theorists about the mechanisms involved in reasoning. Some suggest the existence of general purpose inference rules: subjects are assumed to possess a standard formal logic. Others suppose the existence of domain or context dependent inference rules: knowledge is organized in units called scripts, schemas or schemata which contain rules more or less isomorphic to logic. Finally, a third party assumes the existence of mental models: they claim that people need neither rules nor logic to produce valid inferences. My experiences using LOGO in control technology showed relative stereotypy in the way children were using the programming language. I will try to explain these biases in the light of different reasoning theories. This analysis will come to some conclusions directed to the enhancement of control technology activity LOGO style, and to the modelling of human cognition.

Keywords. Control technology, Learning theory, LOGO, Problem solving, Pupil Learning, Reasoning, Pedagogical robotics.

1 Introduction

In a former study, we introduced control technology activity in some classrooms (see Denis & Sougné 1990). We began by training teachers and then we helped them to introduce this activity with their pupils. Control technology activity lasted for each child two complete days. They were using LOGO language in a constructivist environment with robotics primitives written for controlling the LEGO® interface (Sougné 1989). I took advantage of this opportunity to collect data about children's programming activity. I modified the LOGO editor so that the content of the editor is saved every time children gets out of it. In a former paper (Sougné 1991), I presented these data collected with 12-year-old children. These data were analysed by the means of LOGO-Scan (Sougné 1990), a program written in order to get a picture of LOGO programs. LOGO-Scan produces an inventory of the LOGO instructions used; it gives a logic diagram of the program and it evaluates the degree of chunking optimization of the program. With these data, I analyse the step by step evolution of programming activity.

The analysis of the LOGO programs of one group of children showed that these children were using a narrow LOGO syntax to control robots they constructed:

they used only outputs, they did not use inputs, nor tests, nor other micro world. The analysis of their programs showed that these children built sequential programs made of a chain that sets the electric current on outputs, fixes the current power for a specified time and turns the current off. Programs were controlled by the elapsed time and never by external events. Method of constructing programs began by controlling each machinery separately and then building a super procedure that launched each other procedures. Children seem to be satisfied when it worked, so they did not reorganize programs to eliminate redundant operations. Children did not use any data abstraction (see Abelson & Sussman 1985): neither in building procedures that accept parameters and so could be applied for different purposes, nor in writing programs from the computer point of view. In this paper I will analyze the programming activity observed in the light of reasoning theories.

2 Theories of human reasoning

Knowledge is generally classified in declarative or procedural types, although this distinction has been criticized. For example, Rumelhart & Norman (1988) argue that the context may change what is procedural or declarative. Procedural knowledge specifies how to do things. The knowledge of facts, concepts and relations is qualified as declarative. According to Evans (1989), reasoning skill is a kind of procedural knowledge: on the basis of declarative knowledge (premises, beliefs or observations), humans build deductive or inductive inferences. However, divergences exist amongst theorists about the nature of procedural knowledge that underpin these inferences. These reasoning theories might be separated in three streams of theories: the first one supposes general rules of inferences; the second one supposes domain or context specific rules and the third one denies the existence of rules and proposes the notion of mental model.

2.1 General purpose inference rules

According to authors such as Inhelder & Piaget (1955), Rips (1983), Braine (1978), humans possess a kind of mental logic. This standard formal logic describes and determines the nature of human reasoning. When a subject is faced to a problem, he/she translates its content in an abstract representation upon which inference rules are applied. Once the inference is generated, it is translated back to the real domain.

Piaget does not consider that these general purpose inference rules are innate as Fodor (1980) does, rather they are constructed progressively according to the subject's actions. Piaget (1975) claims that the development of knowledge emerges from the re-equilibration which follows the desequilibration of the subject's cognitive structures. Faced to a problem, subjects try to apply their knowledge by a process that Piaget calls assimilation. Sometimes, assimilation is impossible, a perturbation occurs. A perturbation is all that prevents assimilation to operate. Piaget distinguishes two classes of perturbation that cause desequilibration of cognitive structures:

1. the first comes from object resistance leading to failure or errors,

2. the second comes from lacuna which provoke unsatisfaction of a need; i.e. unavailability of an object, of a precondition necessary for an action to be accomplished or even of a piece of knowledge useful to solve a problem.

A regulation may follow as a reaction to the perturbation. Regulations are always consequent to perturbations, but a perturbation is not always followed by a regulation necessary to attain a new equilibrium. A perturbation may also be followed by:

1. the rehearsal of the same action,

2. the stopping action,

3. renunciation of this type of activity on behalf of an other activity type.

Piaget defines regulation as the rehearsal A' of the action A with A' modified according to the effect of A. Regulation consists in modifying initial action either by modifying the action in increasing negations power (i.e. the action will not be fired again) or by modifying conditions of rule action firing. Generally, a regulation following negative feedback leads to compensation. Two types of compensations are recognised by Piaget (1975):

1. inversion: the perturbation is discarded by inverting the action schema,

2. reciprocity: action schema is changed to make it fit the disturbing element.

Perturbation may also lead to new knowledge by reflective abstraction. This process is a consequence of the subject's experiences. When he/she observes the results of his/her actions on objects, the subject learns, by this experiment, something that he/she did not know formerly.

The process of equilibration of cognitive structures fits fairly well our data. When children have built a program, they test it to see whether it is consistent with their expectations or not. If it does not fit what they wanted, there is a reaction to the perturbation. I observed that:

- children modify some parameters of the actions (modification of the motor speed, modification of the motor rotation duration): the compensation by reciprocity;

- they invert the action (change the rotation of a motor): the compensation by inversion;

- they give up the action (they give up recursion): perturbation followed by the stopping action.

This process is run until the effect of the procedure corresponds to what children wanted. They then try to reach another goal.

These children are far from formal operation stage. They do not use data abstraction; they build program from the robot point of view and not from the computer one and they keep in the program needless operations.

However, even if these children were in the formal operations stage, would they build computer programs in using data abstraction, would they write non-sequential programs controlled by external events and not by time, would they remove needless operations? I do not think so, and we can observe these biases with many novice adult programmers. My own belief is that LOGO does not provide sufficient constraints in itself to force users to write well-analysed programs. All the users want is a program effect consistent with their hope. The cost for building rational programs is too high. If we increase the constraints, we would expect more rational programs and other method of reasoning. The fact is that reasoning is not independent from the context and we cannot take general purpose rules for granted. This hypothesis is supported by the following experiments.

The stream that supports the idea of general purpose inference rules does not really explain content effect on reasoning. The content effect study may be considered as the investigation of relations between procedural and declarative knowledge. According to Piaget's theory, memory (declarative knowledge) and reasoning are two separate systems. Subjects possess logical procedures made of inference rules (Inhelder & Piaget 1955). These abstract reasoning schemata may be applied to every problem belonging to the same logical structure. The fundamental criticism of general purpose inference rule theory (see Nisbett, Fong, Lehman & Cheng 1987, Evans 1989 Johnson-Laird 1991) is based upon what is common to name "content effect in the Wason's selection task". The Wason's selection task (Wason 1966) is a problem in which a set of cards is presented to the subjects. On each of these cards is printed on the one side a letter and on the other side a number. A rule is presented to the subjects. Subjects are asked to turn up the only cards that will help to determine the rule truthfulness. In the example below, presented cards are **a, b, 4,** and **7** and the rule subjects are asked to test whether the rule "If there is a vowel on one side of the card, then there is an even number on the other side" is true or not.

The Wason's selection task

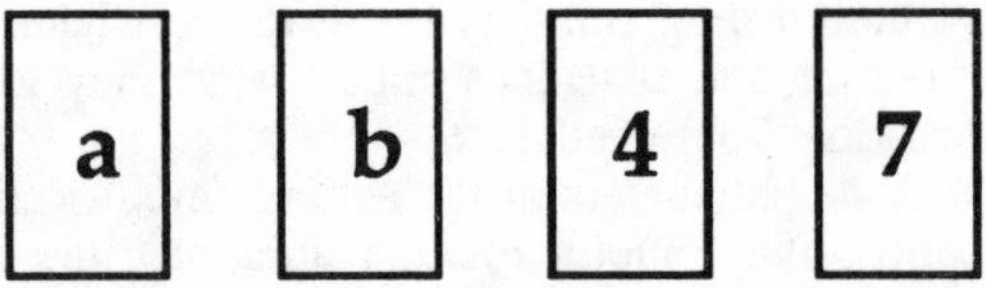

**If there is a vowel on one side then there is an
even number on the other side**

Fig. 1. The Wason's selection task

The correct answer involves turning card **a** up to verify if there is an even number on the other side and the card **7** to verify if there is not a vowel on the other side. Faced to this abstract content problem, a minority of adult subjects performs the right answer. Remember that for Piaget, subjects have the ability of falsifying an hypothesis since they have reached the stage of formal operations. To test the validity of a connective *if ... then ...* the subject will look for a counter-example of p and of not-q (see Beth, Grize, Martin, Matalon, Naess & Piaget 1962, Inhelder & Piaget 1955).

Classical error of the subjects is to turn cards **b** and/or **4** up. They consider the condition as a symmetry even if nothing is said in the rule that behind a consonant there could not be an even number, nor that behind an even number there could not be a consonant. These erroneous answers could have been generated by analogy with some everyday conditions. For instance, if a father says to his son:

"if you drink your soup, you will get a dessert", everybody knows that if the son does not drink his soup, he will not get a dessert. That conclusion is not consistent with logic, but it surely has a functional meaning.

Wason & Shapiro (1971) discovered that with a more concrete content, this problem was better-performed. On one side of the cards was written a name of a town and on the other side a mean of transport. Subjects were informed that each of the cards represented a travel journey of the experimenter. They were asked to verify claims like: *"Every time I go to Manchester, I travel by train"*. Proposed cards were: *Manchester, Leeds, Train, Car*. A significant higher proportion of subjects gave the right answers with this concrete content problem than with the arbitrary abstract content like letters and numbers. Some concrete contents do not facilitate the task (see Manktelow & Evans 1979). According to Pollard & Evans, for a facilitation to occur both the scenario and the content must be adequate. Evans (1989) argues that context and content must be enough coherent in order for the subject to apply actions that are suitable in the real life.

This effect of content might be sufficient to reject general purpose inference rules hypothesis, even if defenders of Piaget will claim that subjects may misinterpret premises because of their implicit knowledge or even if they claim that some people never attain the stage of formal logic. These arguments cannot be reasonably accounted for mistaken responses of college students involved in Wason's selection task experiment. This point has favoured the idea that rules of inference were more specific.

2.2 Domain or content specific rules theories

This approach claims the effect of content on reasoning and postulates rules that one context or domain sensitive (see Grigg & Cox, 1982, Rumelhart, 1980, Cheng & Holyoak, 1985). Rules that a subject will fire will depend of the context and not of the problem logical structure. This should explain the content effect in the Wason's selection task. The explanation of reasoning in terms of specific rules is born with Thorndike (1913). This idea disrupted the old Platonician tradition (Platon claimed that the study of arithmetic and geometry increases reasoning capacities). Note that this belief is still strong in the common sense. The current of specific rules of inference distinguishes logic from reasoning. Newell & Simon (1972) write p. 876: "From the psychologist's point of view, thinking must not be confused with logic because human thinking frequently is not rigorous or correct, does not follow the path of step-by-step deduction – in short, is not usually "logical". Evans (1989) argues that these theories of specific rules of inference explain badly how some humans perform extraordinarily well in face of abstract formulated problems.

Some authors belonging to this approach (Cheng & Holyoak 1985, Rumelhart 1980, Schank 1982) introduced the notion of schema or script. They claim that knowledge is organized in little units called schemas or schemata which are retrieved from memory and adapted to the problem characteristics.

For Holyoak & Thagard (1989), inference as well as analogy should be understood in a pragmatic way, taking into account subjects' goals and intentions.

The pragmatic theory of Holyoak (Cheng & Holyoak 1985, 1989) describes pragmatic reasoning schemas which are a more abstract structures than purely specific rules but more specific than general purpose inference rules. This theory argues that reasoning is neither based on rules independent from the context, nor based on specific experience memory as Grigg & Cox (1982) propose with their *memory cueing hypothesis*. Human would rather use abstract knowledge structures induced by every day life experiences like permission, obligation and causation. These abstract structures of reasoning are called pragmatic reasoning schemas. Cheng & Holyoak (1985) p. 395 define them: "A pragmatic reasoning schema consists of a set of generalized, context-sensitive rules which, unlike purely syntactic rules, are defined in terms of classes of goals (such as taking desirable actions or making predictions about possible future events) and relationships to these goals (such as cause and effect or precondition and allowable action)". As one can see, the purpose, the goal, the prediction and their relations are the organizing structures of a schema triggering.

A pragmatic schema composition

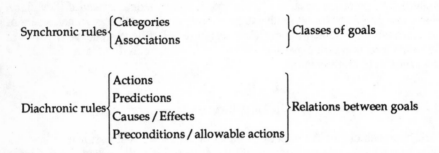

Fig. 2. Synchronic and diachronic rules

Cheng & Holyoak (1985, 1989) explain errors in reasoning by application of a schema which does not correspond to formal logic. All schemas are not facilitating: some provoke non logical responses. An arbitrary statement (as in Wason's selection task) is not connected to past experience and does not invoke a reasoning schema. What subjects do is trying to interpret the problem with their pragmatic reasoning schemas. Failing to do it, they try to generate a conclusion on the base of their knowledge of formal reasoning.

Cheng & Holyoak (1985, 1989) recognize schemas that lead to logical responses. These schemas involve permission: action firing needs a precondition to be satisfied, or obligation: action must be triggered if the precondition is present. Other schemas lead to errors: these schemas involve causation or covariance. Causation has the form of: If <cause> then <effect>. This kind of schemas leads easily to errors since events are often perceived as having a single cause. Therefore problems invoking a causal schema have a chance to generate an inference in the opposite direction: If <evidence> then <conclusion>, just like covariance schema will automatically do. As Piaget argues in Beth & Piaget (1961) p. 195, a subject faced to a complex causality of the form "if p then q", will control this causality by searching for counter-examples "p and not-q" and by verifying if the causality relation is not inverse: "if q then p" in searching for counter-examples "q and not-p". In Wason's selection task, if a subject applies a causation schema, he/she will turn up the card **a** to test if there is not an even number behind a vowel: "p and not-q" and he/she will turn up the card **4** to test if there is not a consonant behind an even number: "q and not-p".

Rules in pragmatic reasoning schemas are created or modified by a process of induction. Holland, Holyoak, Nisbett & Thagard (1986) argue that induction can be reduced mainly in two processes: generalization and specialization.

- There are two types of generalization: condition simplifying or instance based.

1. Generalization by condition simplifying consists in reducing the condition part, for example:
   If x has four wheels and x is on a highway then x is a vehicle.
   ⇒ If x is on a highway then x is a vehicle.

2. Instance based generalization creates a new rule by associating two or more features found in examples:
   The Porsche's are expensive cars.
   The Porsche's are fast.
   The Lamborghini's are expensive cars.
   The Lamborghini's are fast.
   ⇒ If x is an expensive car then x is fast.

- Specialization consist in augmenting the conditions of rules:
   Rolce-Royce's are expensive car and not fast.
   ⇒ If x is an expensive car and x is not tall then x is fast.

One important feature of Holyoak's pragmatic theory is analogy. For Holyoak, analogy implies generalization and specialization and requires category application or formation which is a form of generalization.

The data collected in control technology can be interpreted with Holyoak pragmatic theory in the light of analogy.

Children generate solutions to control the robots by analogy with their real life experience. In Holyoak theory, analogy process is composed of four sub-processes:

1. The retrieval from the base domain to the target domain is caused by the detection of a shared element. These can be wires, lights, motors...

2. The mapping process follows and will generate a solution. The mapping is ruled by three types of constraints:
   - Structural constraints favor single objects or relations mapping.
   Lego® bulbs   ⇔   house lights.
   Interface output ⇒ Lego® bulbs ⇔   switch ⇒ house lights.

   - Semantic constraints favor elements close in meaning.
   Lego® bulbs   ⇔   house lights.

- The pragmatic constraints are the most important and favor functionally relevant mapping according to the subject's goals.

  Turn Lego® bulbs on   ⇔   *switch house lights on.*

  Decrease the motor speed   ⇔   *decrease the speed of an*
                                                      electrical train.

3. The process of solution selection is governed by two criteria:
- There must be synchronic rules between initial states and goal states for both base or target domains. Synchronic rules are a-temporal and concern the categorizations or associations:
  - *Lego® bulbs and house lights are light devices (categorization).*
  - *Lego® constructed cars and electrical trains are motion devices (categorization).*
  - *If an object has a motor then activate the concept speed (association).*
- The base domain must have a diachronic rule between its initial state and its goal. Diachronic rules describe how environment is supposed to change as time goes by, either prediction or action.
  - *If I turn the switch on, then the house light will be on (prediction).*
  - *If the light is red, then stop the electrical train (action).*

4. Learning follows analogy by an induction process. A generalization of the two domains leads to a more abstract schema.

In the first stage, when children make essays of primitives in direct mode (setting outputs on or off, reducing the power ...), they apply a pragmatic schema by analogy with their real world experiences (turn a house light on or off, changing the speed of a motion toy ...). Since it works, generalization of the two domains leads to a more abstract schema. This schema is applied in a second stage when they build their first procedure. Other procedures built successively will be made with the same principle.

Analogy for Holyoak is goal or purpose driven. The pragmatic constraint is the most important and favors functionally relevant mappings. Procedural knowledge involved in a schema is made of rules or heuristics applicable for solving the problem. This interpretation explains why children make use of the primitive that cancels outputs just before setting new outputs. We saw that setting outputs cancels all previously set outputs, hence cancelling an output is useless before setting new outputs. In spite of this reality, children made a wide use of this sequence. According to Holyoak's pragmatic theory, this causation schema is derived from previous experience, probably physics ones, involving electricity in domestic use: turning a light on will not turn any other light off!

Nothing is expected to change since everything goes as expected by children. No specialization follows that would refine pragmatic schemas rules. In conclusion, if one wants to enhance children pragmatic schemas, one must make the children modify their goals, either by increasing external constraints, or by suggesting or even imposing new projects.

Rumelhart's schemata are a bit different and can hardly be called rules. These schemata are an emergent property of PDP (Parallel Distributed Processing) networks. Stimuli get into the system and activate a set of units. These units are interconnected and compose a network. Inputs determine the initial state of the system and the shape of fitness measure. The system transforms itself to reach a stable and optimal state. This state is the product of interaction of many units sets. Some groups of units tend to work together and to inhibit the same units. These coalitions are called by Rumelhart, Smolensky, Mc Clelland & Hinton (1986) schemata. A schema has some properties (Rumelhart 1980 pp. 40-41):

"1. Schemata has variables.

2. Schemata can embed, one within another.

3. Schemata represent knowledge at all levels of abstraction.

4. Schemata represent knowledge rather than definitions.

5. Schemata are active processes.

6. Schemata are recognition devices whose processing is aimed at the evaluation of their goodness of fit to the data being processed."

Rumelhart does not excludes general purpose reasoning skills but specifies that they are a minority in comparison with specific skills. Rumelhart (1980) p. 55 writes: "Most of the reasoning we do apparently *does not* involve the application of general-purpose reasoning skills. Rather, it seems that most of our reasoning ability is tied to particular schemata related to particular bodies of knowledge". He explains content effect in Wason's selection task by the presence of schemata and general strategies of problem solving (i.e. more general schemata). When the problem is not familiar, the subject does not possess a schema to embody the problem, so the only thing the subject can do is applying a general strategy of problem solving. I do not think that Rumelhart's schemata are rules. He defines them more like data structures which represent generic concepts stored in memory. Rumelhart, Smolensky, Mc Clelland & Hinton (1986) claim that schemata are not things, that it cannot exist representational object that is a schema. Schemata rather emerge when we need them, from the interaction of a lot of simpler elements working together.

According to Rumelhart, Smolensky, Mc Clelland & Hinton (1986) a good performance in logic problem solving can be achieved because of some human specific skills. Subjects solve problems less with the mean of logic than with their ability to make a problem fit those they used to solve. Humans are specially skilled in:

1. Pattern matching.

2. Modelling the world: they easily anticipate their action results.

3. Manipulating the environment.

These three qualities explain, according to Rumelhart, Smolensky, Mc Clelland & Hinton (1986), why humans can draw logical conclusions without being logic.

To summarize Rumelhart's views, inputs get into the system, activate a set of units, some of them tend to work together and to inhibit the same units, these coalitions are labelled schemata. The goal acts as an input to the thinking process, and organizes sequence of thought to solve problem. According to Rumelhart (1980), there are three modes of learning:

1. Learning accretion: once we have perceived an event, we can retrieve information about this event.

2. Learning tuning: schemata may evolve to better fit our experience.

3. Schemata restructuring is the creation of new schemata. They are patterned on existing schemata or induced by experience.

The Rumelhart's schemata interpretation of control technology data is rather close to Holyoak's pragmatic theory one. The inputs: the situation as it is perceived and the goal as it is fixed by children activate schemata. Since children have no experience of control technology, they do not have specific schema available that fits these inputs. So a more general strategy is activated. This strategy is probably made of schemata children use for example to turn house light on or off, or to play with toys. These activated schemata are used to create new schemata for the purpose of control technology (schemata restructuring by patterning). These schemata guide the outputs setting, the current power setting, the time controlling and the outputs cancelling. The schemata evolve by a learning tuning process and finally become stable. Since the inputs (the effect of schemata and the goal) do not change, no other schema is activated and the programming activity appears to be stereotyped.

Pragmatic conclusions induced from Rumelhart's theory are the same as the ones derived from Holyoak's theory. In order to enhance reasoning in this control technology activity, the inputs must be changed. There are two ways to modify inputs: either changing the environment constraints or suggesting or may be imposing new goals to children.

2.3. Mental models theories

The argument of Johnson-Laird (Johnson-Laird 1983, Johnson-Laird & Byrne 1991) is that a logical reasoning can be attained without any use of rules nor specific nor general purpose. One can distinguish implicit inferences from explicit inferences. Johnson-Laird (1983) claims that the implicit inference characteristic is that they are generally wrong. He bases his argument on different studies:

Wykes & Johnson-Laird (1977) showed that children have a poor ability to generate implicit inferences to understand certain sentences.

Oakhill (1982) showed that the gap between excellent and mediocre skilled readers consists precisely in their ability to generate implicit inferences.

According to Johnson-Laird (1983), logic is a set of procedures that enables deciding validity of a given inference. However, an explicit inference does not need inference rules nor any formal machinery. Johnson-Laird (1983), concludes that postulating any form of mental logic is useless. An inference system can behave logically without any use of formal rule of inference.

Let's now describe Johnson-Laird's mental model theory. When a subject reasons, what he/she does is testing if the conclusion is true knowing that data are true. Mental models are structural analogues of the world that enable the testing of their truth. When a subject reads the problem premises, he/she already builds mental models to represent possible states of the world which are consistent with available information. The subject constructs a temporary inference based on true propositions of the model. To test the validity of the inference, humans search for counter-examples with the mean of constructing alternative models for which premises remain true, but conclusion becomes false. If no counter-example is found, the inference is taken for granted.

Johnson-Laird (1983) gives an example of the process of inference in a mental model perspective. Two premises are given to the subject:

All   A   are   B
All   C   are   B

1. A model is created:
$a = b = c$
$a = b = c$
   (b)
   (b)   (This model suggests a false conclusion: All A are C)

2. The subject will test this first model. He/She chooses a randomly final item:
$a = b = c$
$a = b = \underline{c}$
   (b)
   (b)

3. If there is an identity between this item and a middle item, that final item is removed:
$a = b = c$
$a = b$
   (b)
   (b)

4. New identities between a free middle item and a final item is searched for:
$a = b = c$
$a = b$
   (b) = c
   (b)

5. The subject tries to verify if the new model is consistent with the premises, and if it is, he/she tests consistency between this new model conclusion and conclusion drawn by the previously constructed model.

6. If this new model conclusion is not consistent with previous one, the previous conclusion is rejected. Then the subject looks for a new conclusion consistent with all valid constructed models (the ones that are consistent with the premises).

7. Go to step 2.

The inference is considered valid if there is no premises interpretation that falsifies the conclusion.

Johnson-Laird explains errors in some logical tasks by the following reasons. Subjects build new models randomly and loose most of their time exploring poor promising conjectures. They also lack of efficient mean to derive reliable conclusions. In addition, the limited capacity of working memory biases the performance. As a matter of facts, alternative models are stored in working memory. Since working memory has a limited capacity (Miller 1956), more the subject has to test alternative models, the more he/she will make mistakes.

According to Evans (1989), this theory can explain belief bias. Subjects may lack motivation to search for counter-examples when a satisfactory conclusion is found to be consistent with the premises. The belief bias appears when subjects start a reasoning task with preliminary beliefs about the conclusion. This bias leads subjects to reject unbelievable but valid conclusions or to accept believable but false conclusion. Subjects focus on conclusion and bring external information to support their decisions. The effort is concentrated to throw discredit on falsifying evidence.

Mental model theory could explain why children control robots with reference to elapsed time rather than with external events. Johnson-Laird (1989) describes the main difference between novices and experts models of physics. Novices construct qualitative models that represent objects in the world and simulate real time processes. Experts build quantitative models that represent abstract relations and properties. This argument explains also why children build sequential programs, and construct a procedure for each machine to control. Experts would rather build recursive programs organised from the computer point of view, with a wide use of arguments that shows data abstraction and controlled by external events (use of inputs).

Johnson-Laird argues that a model can be incomplete and inaccurate but still be useful. This explains why children do not reorganise their programs to eliminate redundant operation.

As a pedagogical consequence, in order to unsettle children current mental models, teachers must incite projects that cannot be solved with the use of children's current models.

3 Conclusions

I do not think that one of these theories supersedes the others, rather I do believe that they explain different facets of reasoning with more or less accuracy.

Piaget's theory, despite the unlikely character of general purpose inference rules, describes a fairly rational and sufficiently precise process of knowledge creation and modification (equilibration of cognitive structures) to be operationalized and implemented on computers. I have implemented a classifier system HCS (see Sougné (in press), Sougné & Blondin 1992) using a new discovery algorithm based on cognitive structure equilibration theory. Another description of modelling Piaget's theory of learning (Schema Mechanism) can be found in Drescher's remarkable book (1991). However, Piaget's cognitive structure equilibration process fails to describe sufficiently how the relation between an action and its effect is reinforced, how generalization occurs and which mechanism presides the choice of a particular reaction to eliminate a perturbation.

Holyoak's theory of pragmatic reasoning is probably more detailed and provides an interesting link with analogy making. This theory has been implemented in a computer programs called PI (Process of Induction); see Thagard & Holyoak (1985), Holland, Holyoak, Nisbett & Thagard (1986).

Rumelhart's schemata provide a framework to link brain neural cells to reasoning behavior. In addition, they provide the advantage to be composed of emergent units which forms a plausible associative structure at different levels of abstraction, a structure that avoids rules rigidity.

Mental models theory takes into account errors in human reasoning. This theory provides a means to understand why people sometimes behave non logically and how they sometimes perform logical behavior without any use of logic. Johnson-Laird (1991) argues that mental models theory is sufficiently operationalized to be modelled on computers. Johnson Laird (1991) provides a description of an algorithm modelling his theory. This algorithm has been implemented and is called PropAi (Propositional inference in Artificial intelligence). Of course, this argument is not sufficient to assess the validity of this theory. What should be done if one wants to model human cognition, is to build a model that is consistent with all knowledge about cognition from psychological data to brain physiology knowledge.

Another criticism of mental model theory relies on the fact that it is so flexible that it should be difficult to set an experiment that could potentially refute it. The way by which mental models theory accounts for errors in reasoning seems to be a bit *ad hoc*, as if Johnson-Laird added to mental models mechanism external processes that bias the performance of the model. For instance, when Oakhill & Johnson-Laird (1985) write that a subject may lack motivation to search for counter-examples. Furthermore, mental models theory does not sufficiently describe how alternatives are generated and how this process is interrupted.

My further research will focus on constructing a model that will be consistent with both psychological evidences and neural framework knowledge. I believe, as Newell (1991) does, that we cannot build useful theories without taking into account all levels of cognition.

References

Abelson, H. & Sussman, G.J. (1985). *Structure and Interpretation of Computer Programs*. MIT Press, Cambridge, Ma.

Beth, E.W. & Piaget, J. (1961). *Epistémologie mathématique et psychologie: essais sur les relations entre la logique formelle et la pensée réelle*. PUF XIV, Paris.

Beth, E.W., Grize, J.B., Martin, R., Matalon, B., Naess, A. & Piaget, J. (1962). *Implication, formalisation et logique naturelle*. PUF XVI, Paris.

Braine, M.D.S. (1978). On the relation between the natural logic of reasoning and standard logic. *Psychological Review*, 85, 1-21.

Cheng, P.W. & Holyoak, K.J. (1985). Pragmatic reasoning schemas. *Cognitive Psychology*, 17, 391-416.

Cheng, P.W. & Holyoak, K.J. (1989). On the natural selection of reasoning theories. *cognition*, 33, 285-313.

Denis, B. & Sougné, J. (1990). *Projet Robotique Itinérante : Rapport final*. Tech. Report. Service de Technologie de l'Éducation, Direction générale de l'enseignement et de la formation.

Drescher, G.L. (1991). *Made-up Minds: A Constructivist Approach to Artificial Intelligence*. MIT Press, Cambridge, Ma.

Evans, J.S.B.T. (1989). *Bias in Human Reasoning: Causes and Consequences*. Lawrence Erlbaum Ass., London, UK.

Fodor, J.A. (1980). Fixation of belief and concept acquisition. In M. Piattelli-Palmarini (Ed.) *Language and Learning: The Debate Between Jean Piaget and Noam Chomsky*. Routledge & Kegan Paul, London, UK.

Grigg, R.A. & Cox, J.R. (1982). The elusive thematic-materials effect in Wason's selection task. *British Journal of Psychology*, 73, 407-420.

Holland, J.H., Holyoak, K.J., Nisbett, R.E. & Thagard, P.R. (1986). *Induction: Processes of Inference, Learning and Discovery*. MIT Press, Cambridge, Ma.

Holyoak, K.J. & Thagard, P. (1989). Analogical mapping by constraint satisfaction. *Cognitive Science*. 13, 295-355.

Inhelder, B. & Piaget, J. (1955). *De la logique de l'enfant à la logique de l'adolescent*. PUF, Paris.

Johnson-Laird, P.N. (1983). *Mental Models: Towards a cognitive science of language, inference, and conciousness*. Cambridge University Press, Cambridge, UK.

Johnson-Laird, P.N. (1989). Mental models. In Posner, M.I. (ed.) *Foundations of Cognitive Science*. MIT Press, Cambridge, Ma.

Johnson-Laird, P.N., & Byrne, M.J. (1991). *Deduction*. Lawrence Erlbaum Ass., London UK.

Manktelow, K. I. & Evans, J.S.B.T. (1979). Facilitation of reasoning by realism: Effect or non-effect? *British Journal of Psychology*, 70, 477-488.

Miller, G.A. (1956). The magical number seven, plus or minus two. *Psychological Review*, Vol. 63, 81-97.

Newell, A. & Simon, H.A. (1972). *Human Problem Solving*. Prentice-Hall, Englewood-Cliffs, NJ.

Newell, A. (1991). *Unified Theories of Cognition*. Harvard University Press, Cambridge, Ma.

Nisbett, R. E., Fong, G.T., Lehman, D.R., & Cheng, P.W. (1987). Teaching reasoning. *Science*, Vol 238, 625-631.

Oakhill, J. (1982). Constructive processes in skilled and less skilled comprehenders' memory for sentences. *British Journal of Psychology*, 73, 13-20.

Oakhill, J. & Johnson-Laird, P.N. (1985). The effect of belief on the spontaneous production of syllogistic conclusions. *Quarterly Journal of Experimental psychology*, 37, 553-570.

Piaget, J. (1975). *L'équilibration des structures cognitives: problème central du développement*. PUF XXXIII, Paris.

Pollard, P. & Evans, J.S.B.T. (1987). On the Relationship between Content and Context Effects in Reasoning. *American Journal of Psychology*, 100, 41-60.

Rips, L. J. (1983). Cognitive Processes in Propositional Reasoning. *Psychological Review*, 90, 38-71.

Rumelhart, D.E. (1980). Schemata: The building blocks of cognition. In Spiro, R.J., Bruce, B.C. & Brewer, W.F. (Eds.) *Theoretical Issues in Reading Comprehension*, Lawrence Erlbaum Ass., Hillsdale, NJ.

Rumelhart, D.E., Smolensky, P., Mc Clelland, J.L. & Hinton, G.E. (1986) Schemata and Sequential Thought Processes in PDP Models. In Mc Clelland, J.L., Rumelhart, D.E. & The PDP Research Group. *Parallel Distributed Processing*. Vol. 2. MIT Press, Cambridge, Ma.

Rumelhart, D.E. & Norman, D.A. (1988). Representation in Memory. In R.C. Atkinson, R.J. Herrnstein, G. Lindzey & R.D. Duncan Luce (Eds.) *Stevens' Handbook of Experimental Psychology*, Vol. 2, Learning and Cognition, 511-587. J. Wiley & Sons, New York.

Schank, R. C. (1982). *Dynamic Memory: A theory of reminding and learning in computers and people*. Cambridge University Press, Cambridge, UK.

Sougné, J. (1989). *Le micro-monde LEGO ®*. Université de Liège, Service de Technologie de l'Education.

Sougné, J. (1990). LOGO-Scan : A Tool Kit To Analyse LOGO Programs. In N. Estes, J. Heene & D. Leclercq (Eds.) *The Seventh International Conference on Technology and Education*. Vol. 2, 313-315. CEP, Edinburgh, UK.

Sougné, J. (1991). Analyse de projets de robotique pédagogique par LOGO-Scan. *Actes du troisième congrès international sur la robotique pédagogique*. CISE, Mexico.

Sougné, J. & Blondin, C. (1992). *Modelling Expertise in Hydrodynamics*. Belgian national incentive program for fundamental research in artificial intelligence, Tech. Report. Université de Liège, SPPS.

Sougné, J. (in press). Modelling Physics Problem Solving with Classifier Systems. In M. Caillot (ed.) *Learning electricity and electronics with advanced educational technology*. NATO ASI Series F, Vol. 115, Springer-Verlag, Berlin.

Thagard, P. & Holyoak, K.J. (1985). Discovering the Wave Theory of Sound: Inductive Inference in the Context of Problem Solving. *Proceedings of the Ninth International Joint Conference on Artificial Intelligence*. Morgan Kaufmann, Palo Alto, Ca.

Thorndike, E. (1913). *The Psychology of Learning*. Mason-Henry, New York.

Wason, P.C. (1966). Reasoning. In B.M. Foss (Ed.), *New Horizons in Psychology 1*. Penguin, Harmandsworth.

Wason, P.C. & Shapiro, D. (1971). Natural and Contrived Experience in a Reasoning Problem. *Quarterly Journal of Experimental Psychology*, 23, 63-71.

Wykes, T. & Johnson-Laird, P.N. (1977). How Do Children Learn the Meaning of Verbs? *Nature*, 268, 326-327.

Cognitive Spectacles

Pierre Nonnon

Laboratory of Robot-Based Pedagogy, Université de Montréal, CP 6128 Montréal H3C3J7, Québec, Canada

Abstract. Cognitive spectacles are a device that allows one to simultaneously observe a physical phenomenon and its graphic representation. This allows the student to grasp the phenomenon under study in both a sensory and symbolic way. This device also gives the student access to induction and to experimental deduction, since, by using a physical phenomenon, he is capable to construct its mathematical model and, conversely, in using this model, he can induce the underlying physical phenomenon.

Keywords. Cognitive tool, Control diagram of variables, Control technology, Deduction, Induction, Knowledge representation, Pedagogical robotics, Science.

1 Introduction

The main objective of this text is to present and discuss the didactic benefits of robot-based pedagogy to the training of students in experimental sciences.

First of all, we will present an inductive approach (action, representation, algebraic formulation) where the student starts by observing the phenomenon, organises the environment in order to induce interactions of variables, and then experiments and constructs models of these interactions in an algebraic manner.

Secondly, we are going to present a new extension of our generic system of science teaching: a deductive approach (algebraic formulation, representation, action) whereby, starting with the model (of the equation), the student can produce the graph as well as induce the corresponding movement.

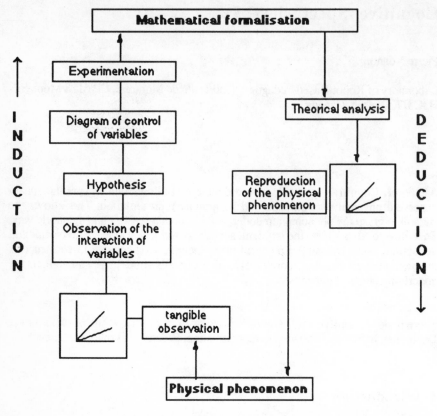

Fig. 1. Generic model of the teaching of sciences: acquisition strategy of experimental induction (Nonnon 1986).

2 The inductive approach

2.1 Action

It takes place in the laboratory where a computer is used mainly to control experiments. It is a slave computer completely dedicated to serving a particular task, in our wordings: a "Robot Appariteur" (see Nonnon et al 1984; 1991) which starts and executes experiments, analyses data and displays them in real time for the learner's benefit. It is through this latter activity of graphic representation of the interaction of variables that we are going to describe the metaphorical idea of the Cognitive Spectacles.

2.2 Representation

Cognitive spectacles are a strategy of acquisition of graphic representation that allows to assimilate this abstract representation to tangible contact with reality. The student, in controlling a toy truck can visualize both the truck's action and the graphic representation of its movement based on time. It is in the simultaneity between action and its representation that he acquires a graphic language of significant coding, allowing him, through the only contact with reality and without resorting to mathematical codification, to interpret and to predict all the movements of his truck. This, we believe, is an interesting educational use of robot-based pedagogy since it allows, as Martial Vivet described to me, to give a double view of the external world, a view which is both tangible and abstract.

This is an effective experimental environment, favourable to the elaboration by the student of formal operative diagrams since he really can abstract information from his own actions on the object (Piaget, 1936). Here the learner's actions lead to the movement of the truck and the dynamic construction of the graph.

It is this dynamic and symbolic representation which should facilitate the connection between the interaction of the physical variables of the real movement and the interaction presented in a graphic form. This activity, by making the graphic representation more signicant, should facilitate its acquisition and make it more available in order to analyse or to predict the truck's movements (Fig. 2, Induction).

2.3 Algebraic formulation

In all our prototypes, after each experiment, the learner is asked to formulate a mathematical expression in order to model the phenomenon under study. He can verify his/her prediction by adjusting, on the outline depicting the interaction of data based on the preceding experiment, a curve drawn from the parameters of the chosen mathematical equation.

3 The deductive approach

In the laboratory, with this new prototype of truck, the student could inversely start by a mathematical formulation, represent it graphically and induce the corresponding movement (Fig. 2, Deduction).

3.1 Algebraic formulation

Here, what remains to be done is again to make use of the preceding equation (in the inductive phase) with its parameters in order to verify whether it generates the same movement and, if it does not, to give a specific value to the equation in order to predict a graph and a new movement.

3.2 Representation

Here the representation will follow the equation, which is a more traditional way to proceed. The student can modify the representation based on the equation's parameters using the graph as a more convenient instrument to predict the movement.

3.3 Action

It will then just be a question of starting the truck in order for it to execute the path laid down by the equation and to verify on the graph that the outline generated by the movement of the truck correctly corresponds to the theoretical outline generated by the equation.

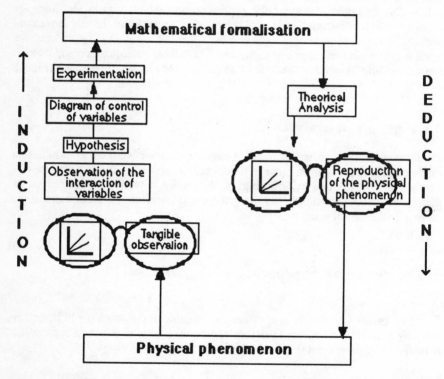

Fig. 2. Illustration of the passage from concrete to abstraction in an inductive approach and conversely, passage from abstaction to concrete in a deductive approach.

In modelling, through an inductive approach, the cognitive spectacles facilitate the connection of the physical phenomenon to its symbolic representation (from concrete to abstraction), (Nonnon, 1986) whereas conversely, in simulation, through a deductive approach, the cognitive spectacles would facilitate this connection between the representation and the phenomenon (from abstraction to concrete).

4 Discussion

To counteract student loss of interest in science and technology, we propose a didactic solution which aims at making science laboratories more attractive with the aim of allowing more students to invole themselves in a creative and constructive manner in knowledge search. Essentially, this solution is based on the practicability of a robotic laboratory which will control experiments and guide students in this approach. The most original contribution of this approach is the fact that it allows a simultaneous visualisation of the action and of its representation. This idea which we have illustrated by the "cognitive spectacles" is a true learning system of abstraction which facilitates the passage from concrete to abstraction in an inductive approach and reciprocally should facilitate the return from the abstract to the concrete in a deductive approach.

In the first phase, we facilitate modelisation of phenomena by the use of graphic representations followed by mathematical ones. We start with a concrete physical phenomenon to go towards the abstract domain of mathematical functions using the graph as a cognitive instrument in order to make the transition from the perceptible to the analytic, i.e. abstraction.

The second phase will allow a reverse in the use of inductive and deductive strategies. The idea behind this being, as is often done, not to polarise the student's activity on one strategy rather than on the other. Here, we can start off again from the deductive analysis of coded knowledge in the form of an algebraic equation, give the equation a graphic form before verifying simultaneously if the movement and the graphic representation really correspond to the deductions that the student made based on its equation and the parameters.

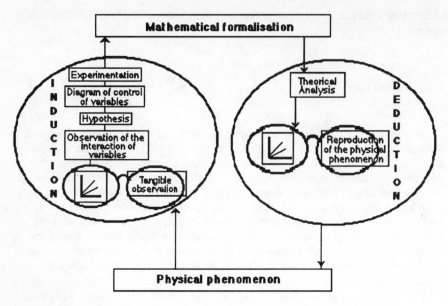

Fig. 3. " The meta-cognitive spectacles"

5 Conclusion

To conclude, we are going to discuss the importance of such a generic system to the development of a scientific spirit. As far as we know, this is a unique system that allows to integrate in a laboratory, within only one prototype, both an inductive and a deductive approach to the teaching of sciences, thus producing a desirable balance between these two scientific approaches. A balance which, as we may recall, has often been upset by the dominance of deductive discourse on the inductive activities carried out in a laboratory. Through this robotic system, we allow the learning, in one laboratory session and on the same phenomenon, of these two complementary approaches. Through these "metacognitive" spectacles, we hope to initiate for the student a complete approach of the acquisition of a scientific spirit. On the other hand, we would like to illustrate in a concrete manner, for the benefit both of the student and the teacher, these two ways of appropriation of scientific knowledge and thus initiate a fruitful reflection on the necessary complementarity of these two approaches.

References

Bachelard G. (1967). La formation de l'esprit scientifique.

Desautels P. (1978). La pensée Formelle. Rapport de recherche inédit. Collège de Rosemont. Montréal, Canada.

Gagné R.M. (1976). The condition of learning. New York: Holt, Rinehart & Winston.

Nonnon et al. (1972). Conditionnement classique et réaction cardiaque chez l'homme. in L. Laurencelle (éd).Bulletin de psychologie. Vol II U.Q.A.M. Montréal.

Nonnon P. (1975). Contingence et discontinuité dans le feedback de l'activité électrocorticale alpha. Thèse de maitrise inédite. Bibliothèeque des sciences, Université du Québec à Montréal.

Nonnon P. Laurencelle, L. (1984). L'appariteur-robot et la pédagogie des disciplines expérimentales. Spectre, Montréal, Canada.

Nonnon P. (1986). Laboratoire d'initiation aux sciences assisté par ordinateur. Publications du Vice-décannat à la recherche, F.S.E, Université de Montréal, Montréal, Canada.

Nonnon P. (1987). Acquisition d'un language graphique de codage par la modélisation en temps réel des données d'expériences. Psychology of mathematics education. Proceedings of the eleventh international conference, Montréal.(pp,228-234).

Nonnon P. (1987). La robotique pédagogique. Le Bus, vol 4 (3), Montréal, Canada.

Piaget J. (1936). La naissance de l'intelligence chez l'enfant, Neuchatel, Delachaux et Niestlé.

Piaget J. (1971). La construction du réel chez l'enfant. Delachaux et Niestlé.

Vivet M. (1983). Logo: Un outil de base pour une formation de base à la robotique. Paper presented at the colloque national: Logo et les enseignements techniques, Le Mans, France.

Language Control Language:
Control Microworlds of the Mind

Mike P. Doyle, Honorary Chairman, LogoS the Logo User Group,
c/o 37 Bright Street, BD23 1QQ Skipton, UK.

Abstract. A theoretical framework is outlined in the context of the teaching of
control technology in elementary schools. The theory uses two important recent
formulations: of information as structure; and of learning as active proposition. It
is suggested that written language be categorised as a technology and the computer
as an extension of human language capability. The notion of technology as
evolution beyond the biologically possible is introduced. Present practice is
analysed from this viewpoint. Disjunctions between the language used and the
activities carried out, both for LOGO Turtles and control technology, are
identified. A focus on mechanism and an intrinsic gender bias is identified. A
programme for future developments which ameliorates these problems and
incorporates biological referents is proposed.

Keywords. Assimilation, Communication, Computer language, Computer
speech, Control technology, Elementary education, Human interface, Knowledge
representation, Learning theory, LOGO, Pupil learning, Pedagogical robotics.

1 Introduction

Control technology involving the use of computers is relatively new. The
endeavour of incorporating computer control technology into the elementary
school curriculum is but a decade old. It is uncertain whether this new curricular
entity is, as yet, adequately conceptualised. The following extract from very recent
advice to teachers suggests that it is not:

... a LOGO procedure for turning two lights on and off in order might be:
  TURNON 1 WAIT 10 TURNOFF 1 WAIT 10 TURNON 2 WAIT 10 TURNOFF 2.
Robot: a mechanical device which can be programmed by the user... NCET (1992)

Teachers, and their advisers it would appear, perceive the computer as a means of
making mechanisms do things; and that their mental metaphor is of something
akin to the Jacquard card. It is the contention of this paper that this conception is
inadequate.

If we are to move forward in the development of control technology as an important curricular element for young children, we need to be very clear about exactly what we intend to teach. The phrase "Control Technology in Elementary Education" contains two pairs of meaning carrying words. The meaning that we assign to these words and their pairings encapsulates our conceptualisation of the field. This then influences the practice of our teaching. We therefore need to be very aware of what we mean by this phrase.

Before moving on to a more formal analysis, let us recall how the computer came to enter the classroom.

In the late 1970s many governments were persuaded that, in some (undetermined) fashion, the curricular use of computers[1] would enhance the economic prospects of the nation.

1.1 Government

In the UK, developments have been almost entirely government directed. Initially control was via the Microelectronics Education Programme and a network of regional Centres. Subsidy has been used to control choice of hardware and supplier. The major thrust has been on the use of applications which mirror those used in industry, e.g. word-processing, databases, spreadsheets, and control. The emphasis has been on the storage, retrieval and manipulation of "information". Computer education in general has largely been guided by civil servants, technologists, and mathematicians.

1.2 Teachers

To some extent, teachers had a different agenda. In the classroom, the assimilation of IT to the pre-existing conceptual framework in education led to a proliferation of programs which added an active element to traditional educational work-cards. These programs, usually mathematically oriented, were often cast in the form of a computer game.

Additionally, a small number of so-called "framework" programs appeared. One of these, "Touch Explorer" (NSNSU, 1992), through its use of the "Concept" overlay keyboard provided a medium through which children could explore and extend a multi-layer data base associated with an illustration[2]. Touch Explorer is one of the few examples of educationalists accommodating to the computer and building a bridge between it and the prior medium of print. It is important in our context because the basis of Touch Explorer is "finding messages".

---

[1] Computer use in schools in the UK is now subsumed under the title "Information Technology", of which it is the major (sole?) component. As we shall see later, this is a somewhat unfortunate choice of nomenclature.

[2] This was achieved by placing the illustration over a flat touch sensitive matrix keyboard.

1.3 Scientists

Computer and cognitive scientists added a third strand. In the beginning, there was a notion that teaching elementary school children to program computers was, of itself, beneficial[3]. In the late 1970s and early 1980s so well did enthusiasts teach BASIC to their scholars that they (too) rapidly became better than most of their teachers.

LOGO entered education within this stream. Launched on the wave of Papert's (1980) "Mindstorms", unrealistic expectations of what might be achieved in terms of cognitive development were raised. Unable to comprehend the nature of and niceties of formal language, educationalists embraced the one concrete referent in LOGO: the turtle (Doyle, 1992a). In the UK, LOGO is now largely synonymous with turtle graphics[4] with a little Control-LOGO thrown in (NCC 1990, 1991).

Part I Theory

Schools in the UK have had a decade to assimilate computers into the classroom. Educationalists now have the maturity and experience to re-state the rationale for the use of computers in school in purely educational terms. This we shall do here through the device of analysing the title of this seminar word by word:

Control   Technology   and   Elementary   Education

There is a temptation to re-order the wording of this title to prioritise the vocabulary. This we shall do in one respect only: in deference to our hosts in Liège we will use the French ordering of noun and adjective for "elementary education".

2 Control

Were the word "control" to be uttered in a school staff-room, it would invariably be taken to mean the management of pupils in the classroom. Utter the same word amongst engineers and it will have, apparently, quite different a meaning. The technology teacher in school will find that s/he has two quite separate concepts of control: in the curriculum, a metaphor of the steam engine governor, or

---

[3] The claimed benefits were not dissimilar to those associated with the teaching of Latin.

[4] In UK Government National Curriculum Orders (the statutory instruments) the phrase "computer language" does not appear in the text and "turtle geometry" is given as an example of the use of co-ordinates in the first quadrant in the Mathematics (1991) Orders.

temperature thermostat; of the pupils, the orderly carrying out of instructions. If we are to comprehend computer control we are in need of a more generalised concept.

2.1 Variety

For the moment, let us abandon the engineer's "feedback loop" notion of control and adopt Ashby's "Law of Requisite Variety". Ashby (1964) illustrates this concept using the device of two players, D and R, engaged in a game. If R's move is unvarying, so that he produces the same move, whatever D's move, then the variety in the outcomes will be as large as the variety in D's moves. If, however, R has two moves, then the variety of outcomes may be reduced to a half, and so on. Thus, if the variety of outcomes is to be reduced, this can only be by an increase in R's variety. Complete control, may be viewed as ensuring a specified outcome. The greater the variety available to the controller, the greater the possibility of determining the outcome. This principle is summed up in the quotation, "In the valley of the blind, the one-eyed man is king".

The notion of variety, as defined by Ashby, is related to the formal concept of information developed by Shannon and Weaver (1949). Blending ideas derived from classical thermodynamics and information theory, Brillouin (1956) proposed that information be considered negative entropy; and hypothesising that objects - the natural and built environment - might be considered "structural" information.

2.2 Information

Stonier (1990) has recently elaborated on these ideas - though he suggests an inverse relationship between entropy and information. This framework he uses to explore the problem of the evolution of increasingly complex structures since the "Big Bang". His proposal, that there is a universal process which converts energy into information (structure), is interesting in that it enables us to view evolution as a cosmic, as well as biological, process.

Whatever the eventual outcome of this approach to theory, we note that it is congruent with the finding that greater complexity provides greater control. A structurally complex organism has greater control over the physical environment than a simpler one: Man can always prevail against the lion at the species level; a stone cannot avoid a falling apple but Newton might have. Control, in the sense developed here, may be seen to be related to structural complexity.

2.3 Anticipation

The engineer's view of control: the damped feedback loop or the instantaneous reaction to sensed change are too specific for our purposes. Control is in the capability to anticipate. Anticipation implies a facility to process messages - data from a distance. It is based on the capability of modelling the structural relationships of the world from the energy patterns which that structure generates.

Thus, a person or machine that can see, hear and feel has more control over the environment than one who cannot see.

2.4 Classroom control

Having made this theoretical detour, let us return to the simpler world of the classroom. A teacher exercises control over the class through the medium of language. There is no direct connection, as in the steam engine governor. If the teacher wishes the scholars to sit down, it suffices simply to say, "Sit down". There is no need to go round each scholar in turn pushing them into their seat. The mechanism of control is communication.

3 Technology

Let us begin with a definition:
Technology is the process of evolution beyond the biologically possible.

3.1 Energy transformation

Stonier's concept of information involves the transformation of energy into structure. Biological organisms can handle energy only within certain limits. Biological evolution must, therefore, have a limit. For evolution to continue beyond the limits set by biological processes higher levels of energy must be transformed. These energy levels are intrinsically destructive of biological organisms. Technology is the means whereby transformation of these destructive energy levels is safely achieved. The test for technology, therefore, is whether the process, or object, is biologically possible.

3.1.1 Tools

Homo sapiens sapiens is generally credited as being a tool-using animal. This is an oversimplification. Analysis of the development of human culture from the cave dwellings of Périgord to the present day shows an exponential growth curve. This growth curve is largely related to a growth in the control and use of ever higher energy levels and the development of tools of ever greater precision. The effect of this on the biosphere in the last century has been considerable. It is as if the population of a new biotechnological organism had exploded.

This process, as Schrödinger (1945) noted, is contrary to the entropy principle. Stonier, as we saw earlier, suggested that it might be explained by taking "information" as an additional physical principle: One which drives the universe to ever more complex levels of structure. Whatever the eventual outcome of this scientific hypothesis building, it opens to us the possibility of proposing that man is more than animal. We suggest that man is essentially technological. That is to say, the quintessential characteristic of humanity is the technology with which human beings surround themselves.

3.1.2 Dangers

We may also note the danger inherent within technology: By going beyond the biologically possible, we have taken control of forces which are biologically destructive. That is to say, our technology is a threat to our biological integrity - an inevitable consequence of the harnessing of energy levels which exceed those of biological processes. Indeed, it is a threat to the biosphere as a whole.

3.1.3 Wheels

If these notions appear extreme, consider the assertion that all our mechanisms are not biologically impossible. Take the wheel. Can it be evolved by biological processes alone? Is the degree of structural precision available to biological processes? We must answer "no"; and accept that the information level (in Stonier's terms) is too low.

3.2 Language

Given this definition of technology, we may examine certain of human developments from a new viewpoint. In particular, we note that the growth of language parallels that of technology. Let us now assert that written language is a technology.

3.2.1 Speech

By definition, speech is a product of biology alone. Ephemeral speech permits of control beyond the senses. It enables imagination. That is, it enables anticipation of events unperceived. Nonetheless, the temporal effects of speech are biologically bound. Oral re-transmission is notoriously imprecise and oral cultural traditions developed a whole variety of devices to stabilise the process.

3.2.2 Writing

Visual representation is intrinsically more stable. Whilst we have pictorial records dating from over 30,000 years ago, we have no records of the language used by these peoples. (Though it was probably highly developed.) Writing did not develop until considerably later. Writing itself has developed along technological lines. From the dichotomy between direct representation and quasi-mathematical symbolic marks seen in the Lascaux cave paintings, through pictogram, ideograph and hieroglyph to the Roman alphabet and numerals; things described and things enumerated remained separate. It was only in the C12 that Hindu mathematicians developed (invented) the zero character and brought the two together. The final act in the development of this unitary notation was the use by Goedel of enumeration in logical proof.

3.2.3 Computers

Turing's notion of a computer was of a machine that could read and write. But the digital computer is more than this: It *is writing*.

This statement should not be found contentious. A digital computer is constructed from electronic components which have language isomorphs in Boolean algebra and number. Like the brain, it processes messages - linearly, though, not across a network. And like the brain it can model itself within itself: Silverman's (1987) version of LOGO implements a set of rules derived form Conway's "Life" algorithm which mirrors on the screen the behaviour of a computer. Furthermore, it has proved relatively easy to develop, within simple computers, a language capability. Indeed, computers have done much to extend our linguistic knowledge. And, in one way we have come full circle: computers can convert text back into an audible form using the same phonetic transcription algorithms used to generate alphabetic writing for spoken languages.

We conclude that digital computer best be considered a language controlled machine. The classic microcomputer arrangement of keyboard, system box and monitor is specifically designed to communicate through language[5]. The task the computer performs, and the manner in which that task is carried out, is entirely dependent upon the program running. For example, the computer does what it has been told to do. It differs from its biological equivalent only in the manner of all machines: by being specific, more powerful, and incomplete.

3.2.4 Babbage

As an aside, we may note that the wheel, as well as the word developed to a high symbolic level: The difference engine designed by Babbage and Baird's spinning disc television are, erroneously, viewed as precursors of our present electronic equivalents; but both were evolutionary cul-de-sacs. The wheel, the quintessential machine, here proved to be less powerful than the word. From this point wheels served merely as the generators of the electrical energy required for digital computers. Babbage's great engine has residual existence in the odometer; Baird's spinning disk in data storage devices.

3.2.5 Biotechnological

Thus, when talking of control technology we are essentially talking about language because language, in its written form, permits of description of the biologically insensible.

Technological man is structurally more complex than biological man and can process information beyond that receivable by the biological senses. So, we see control technology as *action beyond the biologically possible*.

---

[5] Diagrams generated by computer graphics programs are essentially linguistic constructs.

4 Education

Given that we wish to teach control technology, we need a framework within which to conceptualise learning and the learner.

Whilst there are undoubtedly social aspects to learning, in the final analysis learning is an individualised process: The learner learns, yet no one else learns what the learner learns. So, for present purposes, we may concentrate on psychological issues.

4.1 The problem of seeing

Grossberg (1989) has proposed a (very general) mechanism by which learning may take place. His theoretical framework is based on neural net mathematics. Though the specifics of the theory are beyond the scope of this paper, Grossberg's underlying thought experiment is not:

"How is it that with a receptor so noisy as the human eye we can see sharp images, images which remain stable even when movement of the seen or seer is quite rapid?"

Using the known characteristics of the eye and the processing speed of the brain, Grossberg showed, that there is no possibility of stabilising an image using available noise-reduction algorithms. Techniques such as used to enhance images received from, for instance, the Voyager space probe, could not stabilise an image in real time. Without stabilisation in real time vision is impossible. If these approaches were not viable, what alternative might there be? He proposed the following mechanism:

1. The image received at the retina is not, itself, processed. Instead, it is compared with an internally generated image.
2. The processing mechanism is designed to amplify points where the images match and quash those which don't. The process of generating a match continues until the best match is obtained.

He showed that it is possible, within a neural net, to devise a mechanism which will not only perform in this way, but which will stabilise an image in real time[6].

4.2 Learning to see

In this process we see the germs of the mechanism of learning as we know it. For us to see an object the brain must propose that we see it. Thus, the process of learning may be likened to learning to see. Consider the process of learning to weld:

---

[6] This process has aspects in common with data compression techniques.

Initially the novice welder sees only a bright glare around the end of the welding torch. Immediately the pupil brings the flame near a sheet of metal it burns a hole straight through. The teacher will now remark that the objective is to move a pool of molten metal around the surface with the tip of the flame. Given this hint, the novice now finds that the hole takes rather longer to appear; next the pool of metal is seen; and finally, detail on the surface of the pool is perceived as the thin skin of oxide forms and breaks under the guiding influence of the tiny point of flame. The brain has taught the eye see. The physical processes have (perceptibly) slowed down. The hand can move leisurely. The now expert has complete control.

This has all the characteristics of the educational process: Initial "buzzing booming confusion"; a focusing (language) prompt; the slow process of learning to see what was described; finishing with a steady refinement as the learner personally takes control of the learning process and progresses beyond the initial teaching.

4.3 Capability and complexity

As an aside, we note that there is an interesting relationship between processing capacity and capability in a Grossberg system. The efficiency of the system increases with the capability of the organism for predicting the content, and likely direction of change, of an input. It would, at first, appear that the greater the information storage and processing capability available to the organism the greater the likelihood that correct predictions will be made. But this is not necessarily the case. For simple biological organisms in a stable environment it is possible that "pre-wired" expectations will be most efficient. Reduction in the need for pre-wiring would provide an organism with a greater potential to anticipate. It is possible that a critical capacity (or variety) must be reached before language may develop and the bounds of biology be broken.

4.4 Proposing

So far we have used seeing as a metaphor. This is not the best way of conceptualising this system. The brain does not actually "see" anything. What the "minds eye" does is to *propose* what is sensed. It proposes a match to an input. This "propositional" system need not be restricted to matching sensation. Given sufficient capacity, we may conceive of it being used to generate "images" beyond sensation: imaginary entities. In a suitable framework these hypothetical propositions may be tested for their likely consequences in reality. Language provides such a framework. Words such as "if" are the outward expression of this mechanism.

Language also provides a means beyond sensation of communicating the impossible. The means of progressing beyond the biologically possible is not to be found in the tools which are a product of that process, but in the description of the biological process of imagination using language.

4.5 Schools and teaching

An organism which abandons pre-wiring for greater variety needs a system which ensures that fundamental learning is established in safety. The length of post natal dependency is a measure of this. An effective system will both maintain existing learning and provide a platform for going beyond it. We must remember, however, that language is a stable system which embodies a complete culture; hence, new notions take some getting going. Like a pebble tossed into a pond, the ripples may just die away. Embodiment is needed. This has certain practical consequences in the case of major innovations such as IT.

Criticism of schools by certain educationalists (including Papert) notwithstanding, schools and teachers have evolved a very efficient and very stable system for passing on culture, skills and knowledge to groups of children. Consequently, it is unsurprising that IT has been assimilated to education's existing culture. In particular, certain words which differentiate meaning for computer scientists do not do so for teachers. In the case of LOGO, many a schoolteacher will consider LOGO, turtle graphics and turtle geometry to be synonymous. Though this may be painful, intellectually, to those who have passed through this stage, we must not reject these fruits of the assimilation process. It is possible that some of them point to effective strategies towards future goals which those who quickly accommodated to the new ideas might have missed.

5 Elementary

The elementary phase of education establishes that learning prerequisite for independent learning in later phases. In particular, elementary school children learn to use language effectively. They also play.

5.1 Play

The imaginative play of childhood might be thought of as the acting out of "wild Grossberg propositions". Children revel in this activity. The unreal world of childhood is just that. Even reality can be negated. The propositional system, released from realistic constraint by a developing language system, is given full reign. Not only do children go beyond the information given, they often invent the information itself!

5.2 Language development

This is also the major period wherein the world modelled in language is brought into congruence with the perceived world. As this period progresses, control over language increases and pupils' propositions become more realistic. Games, for instance neither have rigid rules nor arbitrary ones; the status of rules as a "legal" framework becomes understood. Conservation over transformation is articulated.

5.3 Pets

We must also note that, both for boys[7] and for girls, the concerns of children in the early elementary stage are more biologically oriented. They focus upon animals, both the family pet and anthropomorphic constructs such as Kermit the Frog and Mrs Tiggywinkle. In this context we note that floor turtles have often been endowed with animistic properties and treated almost as a classroom pet (Mills & Staines 1989). And, equally significantly, that the mechanistic BigTrak has been superseded by the neutral Roamer.

Part II Practice

With the framework outlined above in mind, we may proceed to re-examine past and current practice. Thereafter we may propose progressive developments for which it provides a basis; and finally look a little beyond the presently practicable.

6 The past decade

Two strands in the practice of teaching control technology in education throughout this period may be discerned. These two strands are distinguished by: a) the way in which the computer language chosen expresses control over external devices, and b) by the way those external devices are conceived. These differences, in turn, reflect both the cultures in which they originated and the constraints imposed by first-generation school microcomputers. The strands may be categorised under the headings of Turtle and Control. Under the latter head[8] for the purposes of this discussion, we will focus upon one product only: Technical LEGO®; under the former we shall focus on UK Turtle experiences.

---

[7] Though, consistently, research finds boys in the forefront where things mechanical are involved.

[8] The floor robots, PIP and Roamer (successors to Big Trak) are omitted from this discussion. They are well described elsewhere in this volume. It will merely be asserted, contentiously, that these devices are an unfortunate development. The push-button programming pad each wears on its back is a throw-back to pre-computer technology. It would, for instance, be quite feasible to replace the icons with Chinese characters and the numbers with Roman numerals. These devices only properly belong to technical control, as defined here, when they are enabled to receive and transmit linguistic instructions. This they both can do to a certain extent: PIP can be made to "understand" LOGO; Roamer has its own programming lanugage. Thus, children can write the instructions at a computer and transmit them to the robot which will then carry them out independently. Conversely, they can ask the robot to tell them what instructions has been given by transmitting them to the computer. This "conversational" approach is a more valid use of floor robots.

6.1 Turtle

The Turtle is a ready-made "robot" which is designed to understand certain instructions; in particular, the four geometric and two drawing commands of turtle graphics.

The Edinburgh Turtle, whilst language controlled, was conceptualised as an aid to learning mathematics; hence, as a precision drawing instrument. In its original form it communicated serially with the computer. (The BBC version had a parallel connection, disguised by the use of a round umbilical cable.)

The Valiant Turtle was more obviously language-controlled. It received instructions via an infra-red link. Unfortunately it was its drawing capability, not this mode of communication which captured educationalists attention. This is not surprising given that the communication medium chosen was outside the range of human senses.

Children invested both turtles with animal personalities; often at the behest of their teachers.

6.1.1 Talking turtle

The turtle can be talked to - but cannot reply. This might be remedied, for there is no intrinsic reason which turtles could not communicate in both directions. Indeed, the original MIT Turtle had a bump detector and could send back the message "I touched something".

This is not the major problem with the Turtle. The real problem is the disjunction between the language of turtle geometry used to control it and some very fundamental concepts. If we ask the Turtle to draw a square we say, "repeat 4 [forward 50 right 90]." In both RM LOGO and Turtle Procedure Notation we may also say, "repeat forever [forward 50 right 90]. After a short while the Edinburgh turtle will have twisted its umbilical chord and stopped; the Valiant will continue until its batteries run down. Immediately we see a disjunction between language and reality. Our language implies perpetual motion; our Turtles deny this. The solution to the problem is, of course, to use recursion; see fig. 1.

Unfortunately, in most LOGO implementations tail-recursion is coded as a loop, consequently memory resources will not be use up and the software remains out of congruence with reality.

```
to corner
forward 50 right 90
corner
end
```

Fig.1 A recursive LOGO procedure to draw superimposed squares
to the limit of the workspace.

6.2 LEGO®/LOGO

We take LEGO® Control LOGO as typifying the second tradition. Here a "buffer box"[9] or interface, was connected directly to the computer's parallel output port. The interface, instead of reacting to messages, was umbilically connected to the computer's internal bus. Hence, it and the devices connected to it, were under direct control - more like the limb of an organism that an independent entity.

6.2.1 Bits and bytes

The binary/electronics orientation was further reinforced by the numbering of the connectors on the interface boxes: the first outlet being numbered 0 (zero). Moreover, (UK) Control LOGO displayed a representation of LEGO® interface A on its start-up screen. These developments happened because those driving this aspect of education wanted children to learn about electro-mechanical nuts and bolts.

6.2.2 Switching or telling?

The vocabulary of the software written in the UK for control reflects this direct connection. The UK government sponsored program CONTACT referred directly to the ports, for example SWITCHON 1. This approach was mirrored by the LEGO® Control LOGO developed in the UK, which has primitives such as: TURNON *portlist*, TURNOFF *portlist*, and SENSE? *port*.

In North America, though the interface was identical, the LOGO vocabulary developed for Technical Control LEGO® (under the influence of Papert) retained the "turtle talk" notion. The primitives had the form: TALKTO *portlist* ON, LISTENTO *port* SHOW SENSOR?

Additionally, the LOGO environment, based on LCSI's LOGOWRITER, was more language oriented. Though constrained to parallel connections by economic considerations, the language side helped keep alive the notion of communication.

6.2.3 Constructions

The models designed to be controlled from this interface followed the mainstream mechanical tradition. First came the Buggy (Set 1038) a two motor design which could obey turtle geometry commands; followed by a more comprehensive set (No. 9700) for elementary school use. All the projects in this set were purely electro-mechanical: touch sensors and optosensors, which were used either as switches or counters. Output devices were restricted to motors and lights.

---

[9] Buffer boxes were originally designed to stop teachers from blowing out computer parallel ports and to facilitate the connection of external devices such as motors and lights.

6.2.4 The "knowledge"

The interface box/model arrangement is equally problematical. The disjunction here is not between language and reality but between the language of control and the constructions controlled. Consider: The computer only "knows" about things directly connected to its bus. In the case were are considering, LEGO® TC LOGO, the LOGO language incorporates knowledge about the LEGO® interface card inside the computer and the specific design of the LEGO® interface box (Hence, LOGO will complain if an "impossible" output port is specified).. However, though a LOGO expression to the effect that a motor is connected to port A may be written, there is no way that the computer can verify this.

A child who constructs a model, however, wants the bits that move to do as they are told. That is, the child's mental image is of actions pertaining to the model, not changes of state of the ports. The need to make a motor go by typing either TURNON 1 or talkto 1 on deflects the child's attention away from considering the behaviour of the model to simple switching actions. For secondary phase children such disassociation is not problematical; where children in the elementary phase are asked to do this there must be concern about the educational validity of the activity.

This disjunction, coupled with elementary school teachers' predictable and understandable ignorance of electronics, has made progress in control technology in this phase of education minimal. But this need not be so. The fundamentals of control are not bound to bare electronics. They may be inculcated through devices more amenable to the culture, education and training of those who teach our youngest children.

7 The next steps

If we compare the agenda underlying control technology in schools over the past decade with our analysis of control technology and elementary education we see a mismatch. This mismatch has, largely, come about because a) the first, low powered, school computers were too close to the electronics and b) so were those promoting them. Let us try and free ourselves from the limitations imposed by this now obsolete hardware and begin anew. But, let us temper our progress by a recognition that we must start from where we are; after all evolution, not revolution is our guide.

7.1 Prerequisites

The analysis of Part 2 demonstrated that, in a computer setting, technology is language and control is communication. Education, particularly in the elementary phase, is largely about language and elementary age children are still close to the biological. Technical processes have effects beyond those desired. Indeed, certain side effects are now known to be destructive of the biological substrate which supports us. Furthermore, technical processes, like biological ones, consume

resources. We now begin to appreciate that mechanism alone is insufficient. The biological must be incorporated. Whatever our form our control technology equipment takes, it must:
  a) link to the biological,
  b) embody the concept of resource consumption, and
  c) clearly be controlled communicatively.

7.2 Extending LEGO®/LOGO

Storms of the mind are not on the agenda, simply a series of gradual steps. The construction of new languages and conceptual frameworks is a slow process - there are many old ideas deeply embedded. Let us stick to LEGO® and LOGO.

Consider the following extract from Grey Walter (1961, p. 115) where he describes an aspect of the behaviour of the "tortoise" (M. speculatrix) that was the inspiration for the LOGO Turtle: "The circuit of M. speculatrix is so adjusted that exploration is undertaken in darkness and moderate lights are attractive, whereas bright lights are repulsive. Thus the machine can avoid the fate of the moth in the candle". Surely this is how we would want our youngest children to be thinking.

7.3 Hardware

Grey Walter's Machina speculatrix was a cybernetic "tortoise" capable of exploring its environment. The original MIT turtle was similarly enabled but it was far more capable. Instead of the two "neurones" of the tortoise it had a whole computer brain with which it could communicate via a teletype.

7.3.1 Annaturtle

Doyle (1988) described an experimental development of the Edinburgh turtle which took the original Grey Walter-Papert "cybernetic animal" concept somewhat further. The analogue input capability of the BBC Microcomputer was use to give the turtle "hearing", "vision", "temperature sensing", and a "skin". The days of vacuum tube circuits and gear trains from old gas-meters are long gone. We now have small semiconductor devices small enough to fit inside LEGO® bricks. Hearing was a small microphone and amplifier; vision a light-dependent resistor and temperature a thermistor. The skin was of conductive plastic, to form a touch sensor which was position sensitive. Annaturtle turtle could be made to behave: she could seek or avoid light; find a warm spot; be startled by a loud noise; and test gaps to see if she could pass through them.

7.3.2 Construction

Though a ready-made cybernetic animal has the advantage of removing the disjunction imposed by the interface box, there is no need insist on one. Kits of parts, for slightly older children, have the advantage of being a better focus for the imagination. Let us take the sentiment of Grey Walter and the notions developed in Annaturtle to describe a biologically oriented "LEGO®" kit. We will build bugs, not buggies.

7.3.3 Extending LEGO® Control Lab

The LEGO® company itself has provided us with one step upon the way. They have developed, in the LEGO® Dacta Control Lab, a second generation control interface. Though designed with the traditional secondary science/engineering curriculum in mind, we might borrow it for the elementary school.

This system has a serial interface box which can accept both digital and analogue inputs. The present range of devices is: input - light, temperature, touch, angle; output - motor, lights, buzzer. These are all electro-mechanical in orientation. However, the electronics of the new interface box do allow us to implement sensors with a more biological orientation. We already have eyes (light sensors), and temperature sense. We should now, without too much difficulty, be able to add Annaturtle's "ears" and "skin". This means that, instead of being restricted to modelling industrial robots, we may now devise a "tortoise" which will wake up when the weather gets warm.

7.4 Language

Now that we have exchanged our parallel interface box for a serial one, it is a little more obvious that we are "talking" to it. No longer is there a one-to-one relationship between box and port, eight bits to eight outlets. For, with eight bit serial transmission, we have 256 separate characters with which to communicate. So, we are now sending messages to and from the interface. Hence, the construction `talkto 1 on` is not only preferable to TURNON 1, but is a better description of what takes place.

7.4.1 Fluency

Next, we need to solve the problem of fluency of communication. We need to help our language apprentices to communicate, in messages, with the computer.

A difficulty, to which we are obstinately blind, is that neither teacher nor pupil can type. This makes fluent language communication with the computer almost impossible. One consequence of this has been the excessive use of codes, symbols and abbreviations. For instance, children "talking turtle" are so accustomed to typing the codes: FD, BK, RT, LT, PU, and PD, that they actually articulate the commands as "fud", etc. instead of "forward". Do they, we must ask, then dissociate these codes from the associations that their full forms have with their dictionary homographs?

A decade after the introduction of computers into schools, we must accept that children and teachers are never going to learn to type. We must, therefore, consider alternatives.

An interesting possibility, in the medium term, is that pen input will enable children to write to a computer as they do on paper. This would certainly pace computer use better to children's capabilities; but the technology is unproven.

A graphic user interface with icons is not an option. There is the fundamental objection that "icons" do not make a language system. They have neither the standardisation of representation of the Chinese pictograph or ideogram nor the abstraction of an indicator (Fazzoli, 1987)[10]. In addition, they relate poorly to the nature of the computer as a mechanical language. Recall: a language system must be denumerable before it can be computational. The most appropriate system is the alpha-numeric alphabet. Furthermore, we wish children to express their own constructions in the language - to define new words; for this the alphabet is the only option.

Two solutions to this are presently available. The first makes use of the computer's 'internal knowledge'. The second would be to use a form of intermediate technology, an input device known, in the UK, as an "overlay keyboard".

7.4.2 Active text

One solution to this problem (which may be prototyped within LCSI's LogoWriter) is to use the fact that the computer "knows" what is written on the screen. In this case, the words used for control would be displayed on the screen and the child would select and activate (highlight and click) those required. The cut and paste of word-processing might be used to speed the writing of procedures. Letter-by-letter typing would be needed only once: when typing a new name for a procedure or variable.

7.4.3 Overlay keyboards

The simplest, and most elegant, solution is the overlay keyboard[11] (Doyle 1992b). This is a flat, touch sensitive matrix keyboard. The keys are assigned "messages" in software. Pressing a key on the keyboard delivers the associated message to the computer application. The meaning of the message underlying any key is conveyed to the user by the legend on a paper "overlay" covering the key matrix. (Hence the generic name of "overlay keyboard".)

Keyboard overlays may easily be devised by teachers and pupils. For a specific project, (for instance one based on a standard LEGO® model plan), the overlay might provide all the commands needed to work the model from a pre-prepared set of procedures. For a scratch model, the pupil's own drawing might be used as an overlay; and pressing parts of the picture would call pupil-produced procedures to propel the model.

Overlay techniques may also be used to make explicit the disjunction between the "knowledge available" to the computer and pupil. An overlay which showed the components available for connection to the interface and an illustration of the

---

[10] Icons do, however, betray the perceptions of their author's. A teacher in a local school devised a beautiful MS-Windows turtle icon for LOGO Writer - betraying his (typically British?) perception of LOGO as turtle graphics.

[11] Well known in the UK as the "Concept" keyboard, this input device has, itself, recently become a serial low-power device which will connect to, and be powered from, an RS232 port.

interface itself may be used to "connect", at computer language level, components to the interface connections. Thus, pressing a sequence such as: the legend, "connect"; then a picture of a motor; followed by and an illustrated outlet; would assign a motor to that outlet number in the computer's memory. The computer might then be asked to report how it though the interface box was connected up.

7.4.4 Culture

But we need to take our considerations of language to a far deeper level. Language is not simply a means of communication, it is a carrier of culture. The culture of mechanism is quintessentially encapsulated in the word `repeat` (and the fallacy of perpetual motion in the phrase `repeat forever`). Moreover, language consists of two parts: vocabulary and grammar. In LOGO the linear form of repeat forever [instructionlist] mirrors the mechanical, whilst the syntactical structure of recursion represents the biological. Repeat, a mechanical loop, consumes no resources. It, apparently needs no space to live in. A recursive procedure, on the other hand, replicates itself (arithmetically, fortunately!) and steadily consumes more of the computer's memory. Thus, recursion, a process at the heart of LOGO, expresses the essential language link to biological processes.

7.5 Communication

The serial interface box and the infra-red turtle cut the umbilical to the computer's bus. We are now free of the computer. The moment we move from bytes to messages we cease to be hardware dependent. The interface, now independent, becomes a base for a 'constructible' meta-biological Turtle. We need no longer talk of "ports". We may opt for new terms: the sensors and effectors of biology; the inputs and outputs of computers; or the senses and limbs of people. The model is unimportant. In our "free ranging interface box" we have a structural framework which is capable of evolution in a variety of directions.

7.5.1 How

The means of communication between interface and computer now requires our most careful consideration. Let us focus, again, on young children. If younger children do not see (or hear) the message being transmitted, they have to "take teacher's word for it". (Or note the one to one correspondence between their instructions and the model's actions.) This is unsatisfactory.

The obvious solution to this problem is to take one more step along in our approach to computer language. In the visual display unit (monitor screen) we already have a written language interface to the computer's binary operations. Why not turn this visible text back into "speech"?

7.5.2 Talkers, listeners and audiotext

Recognition of human speech by computers is notoriously unreliable. However, where written instructions are converted into "speech" - or, more accurately, audible text[12] (Doyle, 1986) - using a phonetic speech synthesiser the problem vanishes. Even a very simple speech recognition system will work reliably when receiving audio-text. This opens up the possibility of extremely reliable communication between computer and interface, communication which will remain comprehensible to the children using the system. This approach highlights, once more, the different nature of computer language: technological language which computers and people can share.

7.5.3 Modem chatter

We do not, however, need to restrict ourselves to the slow communication inherent in spoken language. Once the mechanism of communication is understood we may introduce "computer twitter" - binary communication. Most children will now be aware of the sound of facsimile transmission over the public telephone system. The signal processing system required for audio-text may also be used as an acoustic modem. Thus, rapid data transmission from sensors, etc. is also practicable.

7.6 Summary

The freedom language gives us to "broadcast" messages opens up further possibilities. We may conceive of models which incorporate their own receiver and "pay attention" to only those commands for which they have been equipped. In this context, parallels with the notion of a computational "object", which might come to mind, are entirely appropriate. In education we hope to be able to develop from concrete to formal representation. Here we open the possibility of moving from the concrete models of the elementary phase through screen images (active drawings) to formal representations in language at secondary level and above.

8 Discussion

The outline analysis of control technology and its teaching in the elementary school presented here represents a break from present tradition. To the bricks, sticks, wheels and motors of the builder we have added words. Our words are equally technological objects. Like bricks and stick they can be assembled to make new objects; powered by electricity, they can be made to "go", just like a motor.

This unitary notion of technology we have contrasted with biology. We note that through language - the bridging entity between biology and technology - we may model biological behaviour upon a technological substrate. For instance, by conceiving of a light sensor as an eye we may model tropism and a bar-code reader

---

[12] The term "audiotext" has been used to describe the output of intonationless text-to-speech synthesisers by the author.

with equal ease. Focusing on the biological does not degrade the mechanical content; rather it enhances it. The range of constructions (and constructs) open to children is enhanced.

A biological orientation, it may be argued, is essential at our present evolutionary stage. The pupil ecologist must, through play, study a biotechnological word. S/he needs to become comfortable with the notion of bio-technological evolution. With this will come the beginning of an understanding environmental control - control beyond the purely mechanical.

The programme outlined here, though built upon a possibly contentious theoretical base, does not require a massive change in educational practice - just a different viewpoint. Indeed, the more biologically and language oriented programme outlined should be found more amenable by teachers in the elementary phase. The concepts involved and the vehicles for carrying them have greater congruence with the traditions of primary education. The underlying metaphors are more accessible to young children than are the more abstract mechanical notions presently taught. It is hypothesised that the implementation of this programme would reduce, significantly, the gender differences currently to be found in children's reaction to robotics and Information Technology more generally.

References

NCET (1992), Building IT Capability: the Strands of Information Technology, Coventry, UK: NCET

NSNSU (1992), Touch Explorer Plus: TE+ as a TSR for the IBM PC (development paper), Doncaster, UK: RESOURCE/NSNSU

Papert S. (1980), Mindstorms, Brighton, UK: Harvester Press

Doyle M.P. (1986), The Microcomputer as an Educational Medium, Unpublished MPhil Thesis, Manchester University, UK

Doyle (1988), Introducing Annaturtle, LOGOS: Newsletter of the British LOGO User Group, Winter 1988 pp 26-29

Doyle M.P. (1992a), LOGO: little used in the UK after ten years; why?, EUROLOGOS 1 32-35 1992

Doyle M.P. (1992b), Magic Paper: LOGO and the overlay keyboard, EUROLOGOS 1 36-38 1992

NCC (1990), Technology in the National Curriculum: Orders, London: HMSO

NCC (1991), Mathematics in the National Curriculum (1991): Orders, London: HMSO

Mills R. & Staines J. (1989), Turtling Without Tears: Report on the DTI/MESU floor turtle project, Coventry, UK: NCET

Ashby W.R. (1964), An Introduction to Cybernetics, London: Methuen

Shannon C. and Weaver W. (1949), The Mathematical Theory of Communication, Urbana: University of Illinois Press

Brillouin L. (1956), Science and Information Theory, New York: Academic Press

Stonier T. (1990), Information and the Internal Structure of the Universe, London: Springer-Verlag

Silverman B. (1987), The Phantom Fish Tank, Montreal: LCSI

Grossberg S. (1987, 9), The Adaptive Brain Vols. I & II, Amsterdam: North Holland

Grey-Walter W. (1961), The Living Brain, London: Penguin Books

Fazzoli E. (1987), Understanding Chinese Characters, London: William Collins & Son

Meredith G.P. (1966), Instruments of Communication, Oxford: Pergamon Press

What Role Is There for Control Technology in Learning with Computational Expressive Media?

G. Gyftodimos, P. Georgiadis, C. Kynigos

University of Athens, Department of Informatics, Panepistimiopolis, TYPA Buildings, 157 71 Athens, Greece

Abstract. This theoretical paper makes a case for the importance of Control Technology in the development of exploratory software for education. Certain principles concerning both the design and the pedagogy for the use of such software are set against specific properties of Control Technology. It is argued that enactive representation of ideas, understanding of subject-matter principles, use of the notion of variable at different levels of abstraction and notions concerning real world functionalities may well be supported by Control Technology, even to an extent which renders it an essential feature of exploratory systems.

Keywords. Constructionist learning, Control technology, Elementary education, Exploratory software, Expressive media, Learning environment, LISP, LOGO, Programming, Representation.

1 Education and Educational Software: Developing Perspectives

Considering the ITS-based production of educational applications in the last 15 years on the one hand, and the related educational benefit obtained by learners on the other, we are forced to admit that the results are poor compared with what was expected (Eisenstadt et al. 1992). It became clear both that the learning process itself - let alone the process of actually influencing it - is far more complex than originally accounted for by the designers of such software, and that integrating the use of the new technology in educational practice generated further obstacles: not only do we have a very long way to go before computers come anywhere near functioning like human teachers but, even if this is one day achieved (Penrose, 1989), it is unacceptable both by the learner and by the teacher, who both - not without reasons - consider the computer as an annoying intruder obscuring their work.

Initially, the learner's capability of thinking was not perceived as a creative and constructive process (Di Sessa, 1983; Lawler, 1985; Vergnaud, 1987), but as a quantitative gathering of chunks of knowledge; thinking was thus observed as a duty, or as a side-effect to real education which was the acquisition of information through rote learning. The artificial reality of a computer application could neither activate the learner's knowledge construction mechanisms nor support them dynamically even if they were activated. Gradually, even though the process of developing techniques by which machines approximate functions which could be attributed to humans is an enticing issue for A.I., it became clear that such technocentric perspectives (Papert, 1987) were of little relevance to the problematics of human learning at least during general education, at primary and secondary levels (Solloway, 1990).

Educational applications which permit exploration and self-initiated construction of knowledge rather than leading the learner through predefined paths, are much more successfully used provided an appropriate pedagogy is developed and applied. To develop such a pedagogy is a very complex issue, but also one which is integral to any use of technology for educational purposes. We are only recently beginning to fathom the potential of such a pedagogy through a large volume of research describing its implementation in the classroom (Hoyles and Sutherland, 1990; Hoyles and Noss, 1992). In such an eduactional paradigm, the computer application is seen as a tool to develop the way and the breadth of styles by which children can learn a subject matter (Papert, 1981), as well as to learn more about the subject matter itself and at the same time about the expressive power which computer technology provides humans with, an essentially integrated perspective for using computers in a school setting (Polydoridi, 1991; Makrakis, 1988; Budin, 1991; Apple, 1991).

2 Exploratory Software and Control Technology

This paper addresses the role of computer-driven Control Technology within a special kind of computer use, that of the exploratory formulation and expression of ideas by learners and the on-going debugging processes so necessary to constructionist learning (Harel and Papert, 1990; Noss, 1985; Von Glasersfeld E. 1985). What some years ago was known only as LISP-based or LOGO-based programming philosophy (Sinclair and Moon, 1991; Harvey, 1985), is gradually spilling into the design principles of a wider variety of educational exploratory software (DiSessa and Abelson, 1986). One such principle is for the application to allow the user to express ideas in varying forms of representation and by observing the computer feedback to reformulate these ideas and use them at developing levels of abstraction (Hoyles, 1986; Kynigos et al. 1993). The technological developments in the means to represent ideas, the learner-machine interface and the continuous feedback (the potential of which is manifested in applications like Cabri Geometre), have however momentarily subdued the educational importance of using technology to learn about controlling mechanical devices.

Control Technology (CT) enables the learner to control mechanical objects, avoiding the restriction of representing phenomena on the two-dimensional screen. In this case, the means of expressing an idea may be symbolic if mediated through a computer screen, but the outcome is not restricted to a choice between symbols and graphical or iconic representations; it may also be physical, or enactive, to use Bruner's terms (Bruner, 1974).

Moreover, CT is an informatics area which closely combines knowledge on informatics principles with knowledge on the principles of real world events. The interactive "user-computer-actual object" system can thus be observed as a uniform composite entity, with its own rules of operation and evolution, where interaction affects any couple of the components. Amongst a variety of educational applications, those involving CT thus have a special interest, both because CT satisfies specific and well-defined requirements and because it provides a much more meaningful framework to tie in the concrete with the symbolic. In Greece, for example, even though there is very little knowledge about and appreciation of exploratory software (or exploratory pedagogy for that matter), CT has been applied in education, mainly in university physics laboratories.

Since the end of the 1970s, the laboratory experiments for university courses in physics have ceased to be performed manually. In manual experiments, the student participated in a few experiments in the best case and was a remote observer in the worst case, or even only read about the experiment. With the usage of CT, experiments became mechanical constructs with multiple definable parameters, controlled by the computer or through the computer. Experimentation took place faster and was less expensive. We thus have some experimental activity in classrooms even though this is not made pedagogically explicit. For example, it is unavoidable that "error" becomes less a state to be avoided, and more a method to reach the correct state. Sometimes the teachers themselves stimulate the occurence of such errors in the learner's experiments, in order to ensure that the learner does not find the correct state by accident.

The learning mechanism is activated when the human faces a problem meaningful to him/herself and feels the need to find a solution to it (Vergnaud, 1982; Piaget, 1975). CT, by allowing the direct interaction of the learner with the actual object, offers the opportunity of introducing questions concerning the real world and not an artificial representation of reality on the computer screen. In order to use this representation as a space of problems, the learner should be persuaded of its meaningfullness, so that s/he is willing to make a personal effort in order to solve the problems (Hoyles and Noss, 1987). For example, if the robot turtle crashes on the wall, this has implications as an actual event, while the fact that the turtle on the screen reaches the screen margin is only a violation of an artifical constraint. Furthermore, body-syntonic representations of control are no longer the only means for enactive representations, since the controlled robots embody real movement in space. CT may thus be important for early education, not only in learning about CT itself, but also at a deeper level of expressing and representing ideas.

3 Control Technology as a Means for Representing Ideas

Computers are powerful tools for observing the precise consequence of expressing an idea. The educational setting becomes exceptionally rich when feedback is enhanced by the availability of multiple ways of representing these ideas. Children find it particularly difficult to exercise symbolic representations and even more so to use or combine more than one method of representation for the same idea (Mason, 1980; Bruner, 1974). It is very frequent, for example, that in algebra one concrete example is given and then forgotten in favor of symbolic manipulations of formulae where what is actually symbolised gets completely lost by pupils who memorise mechanistic rules. Even when pupils are able to use iconic and symbolic representations in specific situations, they very quickly loose sight of what is symbolised unless they are in a position to jump from one representation method to another. It is thus very important for pupils to have rich opportunity to change amongst methods of representing an idea, and always to be able if required to fall back on to its enactive representation (Mason, 1980).

Computers enable a rich interchange between representation methods belonging to the iconic and the symbolic category. Without CT, it is impossible to use these tools for enactive representations. When acting out the LOGO turtle, children have the opportunity to think in great detail about issues regarding the controlling of a mindless machine and the accuracy of the commands required for such a control (Kynigos, 1989). However, they do not actually experience controlling a machine unless they use a floor turtle.

4 Control Technology as a Means for the Understanding of Principles

Attempting to define the educational subject of CT, we may in general say that the control can concern both the changes (observation of natural changes and modification of their values), and the processes which comprise the actual operation of the object under control, in conditions of mutual interference between the learner and the machine.

This operation can be described in terms of continuous or discrete changes, thus imposing the direct or indirect usage of the notion of variable. Direct usage is performed by naming variables and assigning values to them or retrieving their values, while indirect usage is performed by using primitives, which contain the notion of variable but allow it to be used only as a concept. The description of an operation via straightforward primitives does not only help in understanding the operation; it also allows the description of state changes using terms containing no variables, thus bringing twofold benefits.

Firstly, children who are not old enough to understand the notion of variable can understand and perform parametric operations on actual objects. For instance, children of 9 learn immediately how to use LOGO commands as control commands given by one child and executed by another (Kynigos, 1992b), as part of a game like:

"Now make 2 steps forward, then make 3 steps forward"

thus understanding the notion of "change" in the call of the simple procedure command "forward", without facing the need to express this concept explicitly.

The necessity for formal expressions is also understood as a rule of the game, whereupon the usage of LOGO commands becomes acceptable, like:

FD 5 RT 90 FD 10 RT 90

Thereafter, the detection of repeated similar patterns and the creation of procedures, which replace a sequence of primitive commands, comes as the result of a game of observation and memorising. After this training, when the children are seated in front of the computer for the first time, they simply use the "knowledge of control", which they have already obtained, to control the turtle. This methodology is successfully used in some schools in Athens, in the framework of a long-term project, which studies the possibilities of modern technology usage in the first level of education (Kynigos, 1992). So, the children use various conceptual quantities without being aware of using variables.

Secondly, the "variable" is conceived as a name necessary to identify a "conceptual object", so that it can be implemented and used. The notion of variable can be introduced even to children of 10 by using LOGO, by considering quantities which are known to have some straightforward meaning for the child. Such a quantity cannot be, for example, the size of the quadrant, because it lacks a clear criterion of what is measured. On the other hand, a simple reference to the variable of a primitive, such as in the procedure " TO GO :X I FD :X I END ", is not appropriate either, because the goal, which is to understand the notion of renaming the command, is achieved in a too formal way.

An appropriate quantity for this purpose could be the duration of a music tone: when the children learn the primitive TONE and experiment with different values of the primitive's parameters, they understand very easily the role of the parameters. Thereafter, when they learn how a music note is created as a procedure, for example " TO LA I TONE 440 20 I END " , they immediately want to know the specific values of the frequence parameter in order to create the other notes of the musical scale by themselves. However, children quickly realise that the only way they have for making the computer play a note in different durations is by creating different procedures; this is a boresome work, so that the introduction of the notion of variable is accepted with enthusiasm: the grouping of many procedures in one, corresponds to the replacement of many manual operations by one.

The sound manipulation in LOGO is a CT operation, because it concerns the control of an actual object (the sound), which is conceived directly and is controlled via the computer. Given that no specialised hardware is necessary, sound manipulation is an appropriate means for presenting CT to children, with or without usage of variables, by allowing the creation of complex procedures and by referencing composite objects (repeated phrases, variations ...). Furthermore, the combination of sounds with visual shapes enables the categorisation of activities in a straightforward and impressive way: for example, the description of world evolving over time as a "musical shape".

The control over a specific operation has the meaning of interactions with it, which may take the form of simple informative questioning, of invocation of changes or of redirection of the operation's evolution, where all forms of interaction allow the user to deduce knowledge on the rules of behaviour of the object, and to foresee events of the real world in which it appears. This interaction is expressed via access to variables, irrespectively of whether these variables are named or not. Therefore, it is very useful to have the operation analysed as a number of procedures, and to specify the interdependencies among them, preferably in a "forest" structure.

The LOGO environment is sufficiently expressive for describing a world and its operations. Although LOGO is not the only environment appropriate to the purpose, some of the LOGO features are available only in very complicated programming environments, like LISP, or in more open programming environments, to the point of being difficult to use, like PROLOG. Such features of LOGO are not only the primitives of turtle graphics, but also the primitives of list processing and property lists. Moreover, equally important features are the automated replacement of a variable with another, as well as the binding of a variable to an object, probably dependent on other objects which in turn could contain other variables as expressed by lambda calculus rules (Abelson and Sussman, 1985). These features are very important CT: the real world has its own internal structure, which can be directly controlled only if it can be exactly mapped on the computer. The LOGO features just mentioned enable such mappings in a straightforward way.

5 Complexity of Control

There is a limit to the potential complexity of the manipulations of the object under control. This limit is fixed more by the capacities of the available material and by the user's mental abilities and skills, than by the theoretical requirements of the experiment under construction: an experiment is a world to explore. So, it must be presented to the learner not only as a medium for conceiving a specific rule, but rather as a space where the learner is called to find out what other events take place (or could take place), how his/her observations are applicable to the same or to another similar construction, how the control over space is improved, how objects can be grouped together, and which special properties can be further identified. Therefore, the experiment initially designed must be able to evolve, to be extended, generalized and/or specialized, without the need to interrupt the current operation of the system. In this way all actions of all possible changes are perceived by the learner as belonging to the same world and they are evaluated as actions of conceptual abstraction / concretization.

Thus, the first priority for the selection of CT software and hardware in an educational environment should be given to the ability it has to satisfy not only the current needs of the curriculum, not only each possible requirement of the learner and each idea of the teacher for further development of the system, but also each future requirement for further complexity. It is reasonable to claim that the above requirements are satisfied by any programming environment conforming to a LISP-based philisophy (as it is the case of LOGO), so that, for example, when the learner is no longer satisfied by the features of this environment, then s/he will be already prepared for the next step (as it is the case of LISP).

6 The Different Notions of "Variable" in Control Technology

In the conventional usage of CT in the laboratory for controlling some multiparametric construction, the variables under control play some simple but distinguishable roles: there are variables expressing continuous or discrete quantities, logic values or higher levels of taxonomy, indicating case selection for a specific action or the whole experiment. Inevitably, there is also a different class of variables, mainly those concerning the management of the simulated world (computer and mechanical construction). These variables are not related with the experiment itself as it is considered to take place in the real world.

The conceptual isolation of the latter class is imperative for the educational success of the experiment: we must oversee them. This causes an important didactic problem, which is more acute when the educational level in which CT is used is low. However, the problem can be solved because children are accustomed to simulations of reality in their games: there, the isolation of some parameters solely concerning the world of simulation is obvious.

However, since it is not certain that the learner is aware of such differences of variables in the experiment, it must be ensured in advance that s/he understands the following:
- the entities directly related to the real world s/he is studying,
- the dissimilarities between the actual object construction and the real world entity it represents, which must be overseen for the needs of the experiment
- values that may not be modified, because they are constants of the real world entity or because they are beyond the capacities of the implemented construction
- the range of fault tolerance in the real world entity as compared to the fault tolerance of the actual implementation
- the reasons for performing specific actions, distinguishing between those allowed or needed on the real entity and those on its implementation

The explanations of the teacher are of little help if the learner has no direct understanding of the functionality of the real entity. Thus, whenever a simulation construction is used, the teacher must verify in advance that the learner is aware of the rules and the states of the simulated world.

7 The Meaning of World Functionality in Control Technology

Contrary to full computer simulation of a complex object, CT allows direct observation of the object, even if its construction is oversimplified, because the construction is still operating in a natural way. For example, a moving component has a size, volume, weight and position in space, and events take place in real time. So, the learner using CT becomes aware of the object's actual properties in a direct way and not because the teacher assured him/her that the object simulated by the computer does have the actual properties.

However, despite the directness of observation offered by CT, the confusion between the supposed real functionality and the functionality of actual object in the experiment is still possible to occur in the learner's mind. This problem is more acute than the previously mentioned problem concerning the distinction among the classes of variables: this distinction can at least be enforced by a formal representation, probably in an advanced logic system like predicate calculus or in lamda calculus, while the confusion between the different functionalities can only be overcome if the learner understands the ontological perspective of both the real entity and the actual object. In a classic computer-simulated representation of a real world entity, the implementation requires that the entity is fully analysed as a set of well-defined states and rules/methods. Quite differently, for a CT representation it is sufficient that only some of the real world parameters are under control. Moreover, it is possible that these parameters are deliberately selected to present the desired functionality, when constructing the actual object. So, it is not necessary that every aspect of the real entity be mapped on the actual object.

In short terms, both the real world and its simulation have their own sets of rules. Some rules are common (e.g. gravity, inertia), while others are overlapping (e.g. a real object has only some of the properties of the real entity it represents). From that point of view, systems like the "greenhouse" of LEGO-LOGO have the advantage of being real worlds themselves rather than simulations.

8 Programming in a Control Technology Environment

Programming in the CT environment concerns both the establishment of the appropriate premises for the experiment by the teacher or by experienced programmers, and the interaction of the learner trying to describe his/her ideas, which may be either facts to be fed to the computer or rules, i.e. methods to be applied on the facts.

Among the merits of LOGO, the ease of programming should be acknowledged. CT applications can be implemented in many other programming languages (Pascal, C++, ...) producing technically more efficient code. However, in such a case, the user must have a sufficient level of programming expertise, and is enforced to express his/her interventions via complex language constructs which blur the concept materialised by the intervention.

A CT-construction which is fully LOGO-controlled, i.e. where full LOGO programming is allowed as opposed to the simple grouping of primitive LOGO commands, enables the creation of procedures, which correspond to real simple or complex activities. Such procedures can even refer to more generalised objects (object classes represented as lists), to object properties, to chains of relationships etc, or even be observed as generalised objects themselves.

Obviously, not all these features are necessary to learners of any age. Their absence, though, places narrow limits both on the functionality of the applications and on the range of ages of learners, for which they are appropriate. Furthermore, due to the lack of widely accepted standards, the knowledge obtained by the learner in some level of education finds no continuation in subsequent levels, if the learner is forced to learn a new application each time a new conceptual development is reached.

Although it is rather difficult to describe a hierarchy with property inheritance in LOGO, the child considers it obvious that "since the canary is a bird and since birds have wings, the canary has wings" and expects the computer to make such "simple deductions'". So, it is necessary to organise the LOGO environment on a per case basis by attaching the necessary rules to the application program, so that such simple deductions of the learner are embedded into the environment.

9 Control by Means of Property Manipulation

Understanding the properties of an object, the subparts from which it is composed and their specific function and the behaviour of that object within a given environment requires the ability to control the object by changing values of properties of the object itself or those of the surrounding environment and appreciate the presice effect these changes have in the objects behaviour. The design requirements for such a device and for the controlling inteface are complex. On the one hand, the device itself has to have a realistic appearance and behaviour. For example, it is educationally unacceptable for a floor turtle not to rotate the exact amount commanded by the user because one wheel finds less resistance to turn than the other. The manufacturer may well perceive this as a trivial problem of turning a screw to loosen a hinge. The learner however may confuse the meaning of rotation, the quantity of a degree and the notion of control all together, since the device did not behave as commanded by the computer. On the other hand, the symbolic representation of the commanding primitives and the primitives themselves must have a tight connection with the learning target and with off-computer symbolisations controlling such a device.

It is therefore important for a computer environment to support the creation of computer objects, the links between those objects and their properties and the tight connection between the values of the properties with the objects real situation and behaviour. A LOGO-based CT environment incorporating input facilities for the appropriate sensors does support the above: LOGO's property primitives (such as PPROP, GPROP), provide the means to build higher order primitive procedures embedding the required notional links between property and behaviour. The educational power of such an environment is that the learner may look inside these primitives in order to understand the effect of their execution, change parameters and most importantly, build his/ her own without having to learn to use any additional technology.

10 Conclusions

The argumentation built in this paper for the role of CT in learning with exploratory software does not imply that CT is free of unsolved problems. For instance, it is true that using a computer under CT enables the learner to practise control over actual situations rather than over a poor simulation of the real world on a computer screen. It could be claimed, though, that the actual object under control is still a poor representation of the real world entity, and a rather complicated one, on which not all ways of control conceived by the learner are applicable, easy to perform or feasible at all.

When the learner works in a "world" supervised by automated mechanical components, for which s/he only knows the input and output information, and when s/he is called to solve high level problems, while lower level problems are tackled by the computer itself, the learner may learn faster and obtain broader knowledge, but s/he may also loose the opportunity of obtaining the deep specialised knowledge achieved by lengthy intensive occupation with the problem, resulting in its analysis down to elementary primitives.

Such problems obtain particular importance as the CT applications are extended to cover more and more educational subjects, and as CT is used in gradually lower education levels. In order to solve them, CT should be placed within a global long-term didactic program, to be attended for long by the same learner. It is only within such a program that we can ensure that there is no subject about which the learner has never learned, because it was always handled by a computer.

There are indications that the educational gains from the usage of CT are broader than the objectives of each occasional application. Informal reports of students who have worked in CT environments during their studies, indicate both adaptability to new requirements and a skill to exploit the variety of new information they receive. The enriched interplay amongst methods of representing ideas, the analytical thinking and expression of ideas, the ability to generalise notions from observing specific outcomes from combining feedback from both the computer and the control device and to express these generalisations via programming are some learning features for the encouragement of which CT may be a powerful tool.

Acknowledgements

We would like to thank Dr. M. Spiliopoulou for translating and commenting on an earlier draft of this paper.

The study presented in this paper is one of the diverse activities of a permanent seminar in Computers in Education, held at the Department of Informatics, University of Athens, and directed by Prof. G. Philokyprou.

References

Abelson, H. and Sussman, J. (1985) *Structure and Interpretation of Computer Programs.* MIT Press, Cambridge MA.

Apple, M. W. (1991) *The New Technology: Is It Part Of the Solution Or Part of The Problem In Education?* Computers in the Schools, vol 8 (1, 2, 3), pp. 59-81, Haworth Press.

Bruner, J. S. (1974) *Beyond the Information Given,* Allen and Unwin, London.

Budin, H. R. (1991) *Technology and the Teacher's Role.* Computers in the Schools, vol. 8 (1/ 2/ 3), pp 15-26, Haworth Press.

DiSessa, A. (1983) *Phenomenology and the Evolution of Intuition.* In: Mental Models, (Gentner D & Stevens A, Eds), pp. 15-33, Lawrence Erlbaum, Hillsdale, NJ.

DiSessa, A. and Abelson, A. (1986) *BOXER: A Reconstructible Computational Medium.* Communications of the ACM, 29 (9), 859 - 868.

Eisenstadt, M., Price B.A. and Dominique, J. (1982) *Software Visualisaton: Redressing ITS Fallacies.* In: Cognitive Models and Intelligent Environments for Learning Programming (E. Lemut, B. du Boulay, G. Dettori, eds.), NATO ASI Series F, Vol. 111, Springer-Verlag, Berlin.

Harel, I. and Papert, S. (1990) *Software Design as a Learning Environment.* Interactive Learning Environments, 1, 1 - 32.

Harvey, B. (1985) *Computer Science Logo Style.* Vols 1,2,3. MIT Press, Cambridge MA.

Hoyles, C. (1986) *Scaling a Mountain - A Study of the Use, Discrimination and Generalisation of Some Mathematical Concepts in a Logo Environment.* European Journal of Psychology of Education, 1 (2), 111-126.

Hoyles, C. and Noss, R. (1987) *Children Working in A Structured Logo Environment: From Doing to Understanding.* Recherches en Didactiques de Mathematiques, 8 (12), 131-174.

Hoyles, C. and Noss, R. (1992) *A Pedagogy for Mathematical Microworlds.* Educational Studies in Mathematics, 23, 31-57.

Hoyles, C. and Sutherland, R. (1990) *Logo Mathematics in the Classroom.* Routledge, London.

Kontogiannopoulou-Polidoridi, G. (1991) *The Educational and Social Dimensions of the Use of the New Technologies in the School,* paper written in Greek in Synchrona Themata, vol. 46-47, Dec. 1991, pp. 77-93.

Kynigos, C. (1989) *Integrating Technology into the Culture of a Primary School.* Proceedings of Eurologo '89 (Schuyten G and Valcke M, Eds), pp. 457-468. EDIF, Gent.

Kynigos, C. (1992) *Insights into Pupils' and Teachers' Activities in Pupil - Controlled Problem - Solving Situations: A Longitudinally Developing Use for Programming by All in a Primary School.* In: Mathematical Problem Solving and New Information Technologies: Research in Contexts of Practice (J.P. Ponte et al., eds), NATO ASI Series F, Vol. 89. Springer-Verlag, Berlin.

Kynigos, C. (1992b) *The Turtle Metaphor as a Tool for Children Doing Geometry.* In: Learning Logo and Mathematics. (C. Hoyles and R. Noss Eds), MIT Press, Cambridge MA.

Kynigos, C., Gyftodimos, G. and Georgiadis, P. (1993) *Empowering a Society of Future Users of Information Technology: A Longitutional Study of Application in Early Education.* To appear in European Iournal of Information Systems.

Lawler, R. W. (1985) *Computer Experience and Cognitive Development. A Child's Learning in a Computer Culture.* Ellis Horwood, Chichester, UK.

Makrakis, G. (1988) *Computers in Education, Studies in International and Comparative Education.* Stockholm Institute of International Education.

Mason, J. (1980) *When is a Symbol Symbolic?* For the Learning of Mathematics, 2, 8-12.

Noss, R. (1985) *Creating a Mathematical Environment through Programming: A Study of Young Children Learning Logo.* PhD. Thesis, published by University of London Institute of Education.

Noss, R. and Hoyles, C. (1991) *Looking Back and Looking Forward.* In: Learning Logo and Mathematics. (C. Hoyles and R. Noss Eds). MIT Press, Cambridge MA.

Papert, S. (1981) *Computers and Computer Cultures.* Creative Computing, 7, 82-92.

Papert, S. (1987) *Computer Criticism versus Technocentric Thinking.* Educational Researcher, 16 (1), 22-30.

Penrose, R. (1989) *The Emperor's New Mind. Concerning Computers, Minds and the Laws of Physics.* Oxford University Press.

Piaget, J. (1975) *L' équilibration des Structures Cognitives* (EEG XXXIII). Paris: P.U.F.

Sinclair, K. and Moon, D. (1991) *The Philosophy of LISP.* Communications of the ACM, 34 (9), 41-47.

Solloway, E. (1990) *Quick, Where do the Computers Go?* Communications of the ACM, 34 (2), 29-33.

Vergnaud, G. (1987) *About Constructivism.* Proceedings of the Eleventh International Conference for the Psychology of Mathematics Education, 42-55. (Bergeron J, Herscovics N and Kieran C, Eds), Montreal.

Von Glasersfeld, E. (1985) *Representation and Deduction.* Proceedings of the Ninth International Conference for the Psychology of Mathematics Education, 484-489. (Streefland L, Ed.), Noordwijkerhout, The Netherlands.

Weir, S. (1987) *Cultivating Minds: A Logo Casebook.* Harper and Row.

2. Experiments and Case Studies

2.1 Course Contents

Robotics and Telecommunication (Experience of the Ecole Active de Malagnou in Geneva Within the Framework of a Computerized Educational Environment)

Jean-Claude Brès

Ecole Active de Malagnou, 39b, Route de Malagnou, CH-1208 Genève, Switzerland

Abstract. Since 1983, the Ecole Active de Malagnou has been using computers in the classroom. Following a year of research and travel to observe various experiences in integrating computers in classrooms in different countries, this school implemented a computerized educational environment (EPI - Environement Pédagogique Informatisé) in 1987. The EPI project has five avenues (CAI, Desktop publishing, Robotics, Telematics, Creativity).

Students work on global projects, in other words, on themes that require observation, visits, and crossing disciplinary boundaries, and that result in the creation of programmed robotic models or simulations (LEGO-LOGO).

The establishing of exchanges using telematics (modem and networks, fax, minitel...) with other classes is now our goal.

Pangea (*Pangée*) is the name given to a new robotic and telematic project which has children from all continents communicating on the theme of ecology. The project started during the last school year and is meant to last several school years giving time to the network to spread over the continents.

Keywords. Communication, Constructivism, Control technology, Elementary education, LOGO, Longitudinal study, Problem solving, Project driven learning, Pupil learning, Pedagogical robotics, Socialisation, Telecommunication.

1 The Roots of the Ecole Active de Malagnou

The Ecole Active de Malagnou was founded in 1972 by a group of educators in Geneva. At the heart of the group was Professor Michael Huberman (now at Harvard) who at the time was the President of the Science of Education Faculty at the University of Geneva. The other members of the group were:
- Claude Ferrière (son of Geneva educator Adolf Ferrière, author of the book "The Active School"),
- Robert Hacco (Foundation for Modern Education),
- Laurie Lamartine (psychologist), and
- Freddi Stauffer (a parent).

This school is part of the movement of "active" schools, whose roots are found in the theories of autostructuration of thought (Montessori, Decroly, Claparède, Dewey, Freinet). Created in Geneva, and growing in a constant relationship with the university, the Ecole Active de Malagnou is, of course, seeped in the Piagetian approach. The school has always sought to remain alive. Through contacts with the French Association of New Education (GFEN— Groupe Français d'éducation nouvelle) or the French Reading Association (AFL—Association Française pour la lecture), for example, or through its own activities and research, the Ecole Active of Malagnou continues to evolve its theory, and its practice, towards a pedagogy grounded in the theory of the interstructuration of the subject and the object, as described by, among others, Louis Not, in a book entitled *Pedagogies of Knowledge* (Les pédagogies de la connaissance).

I would like to quote Not, whom I have loosely translated:

"It is interstructuration of the subject and the object... By taking into account the genetic dimension, this method introduces, as essential, the initiative and the activity of the subject in the construction of his/her knowledge, refuting any process of transmission of knowledge, or of co-action. Rather, by exploiting the structuring effects of cultural content, saving the student from a slow and total rediscovery, and rendering useless any recourse to camouflage by a hidden directivity."

2 Blowing Up the School

Some innovative educators, some researchers or theoreticians have dreamed of "blowing up" the school, opening it up to the outside world (the students' families, the city, other schools, current events worldwide, interculturalism). Some students are bored to death at their school desks in an overly traditional school; they too have dreamed of blowing up the school. I was one of them.

Well, it has been done. The school has been blown sky-high. The explosion has begun on several levels.

2.1 Material Potential

- Today, a considerable number of students have a television set and even a video recorder at home.
- It is at home that one finds the most computers, not at school; the situation is the same for tape recorders, books in general, encyclopedias, etc. (also true outside of the industrialized countries, for instance, recent statistics on Latin America, presented by Professor A. Battro at the 3rd LOGO Congress in Petropolis, Brazil.)

2.2 Technological Innovations

A number of technological innovations are in the process of radically changing facts in the area of communication and computer science in general, and through that, are changing education.

Among these new technologies, a few seem to me to be particularly important.
- Telematic communication networks, for example, Bitnet, which connects educators and university researchers throughout the world. This network is beginning to be used by teachers and their classes (Kidnet, or Pangea, for example) or the AGE network (Apple Global Education).
- The introduction of wireless computerized communications (wireless data communications or mobile wireless computer network such as "Project Wireless Coyote by ACOT (Apple Classroom of Tomorrow)" — use of amateur radio frequencies, with the radio packet system (for example the work of Lea Fagundes, who works with street children in Brazilia); the use of communications satellites.
- Last, but not least, the introduction on the marketplace of the interactive CD Walkman.

These means taken together (of which some are quite inexpensive: radio communications, CDs) allows us to imagine the opening up of each school, a window onto the world of today, the entire world, as it is right at this moment. This is no longer just a dream.

3 Computerized Educational Environment (Environnement Pedagogique Informatisé - EPI)

Since 1983, the Ecole Active de Malagnou has been using computers in the classroom. Following a year of research and travel to observe various experiences in integrating computers in classrooms in Switzerland, France, Belgium, the United States and Québec, the Ecole Active de Malagnou in Geneva implemented a computerized educational environment (EPI) in 1987.

The *EPI* was created as an educational project whose objective is to apply new technologies in areas where they are particularly effective, and where they participate in the creation of a truly active pedagogy.

The *EPI project has five avenues:*
- Computer Assisted Instruction
- Desktop Publishing
- Robotics
- Telematics
- Creativity

3.1 CAI (Computer Assisted Instruction)

As with most experiments integrating computers in the classroom, CAI remains relatively limited in the amount of time that students and teachers devote to it.

Few software tools are flexible and used regularly (some exceptions exist such as Elmo within the framework of many reading workshops, structured according to the approach of the AFL (French Reading Association); or Cabri-géomètre and Cabri-collage (for geometry.)

3.2 Desktop Publishing

The name might be a bit pretentious, but essentially, what it really means is using word processors, drawing software, and in the case of publishing a class newspaper, a page layout program. In any case, it is a use of the computer that teachers and students favor: the students publish short stories, novels, a weekly paper with a regular production staff: they also publish research papers based on historic, scientific, and documentary research.

3.3 Creativity

Two domains have been the object of important projects: visual arts and music. The relationship between the video image and the computer was explored in a project called "Portrait". Under the responsibility of Pierre Dunaud, this project integrated digitalized video images, drawing programs and word processing, parallel to traditional graphic design tools (felt pens, paper, paint...).

Music was integrated in the framework of a workshop led by Alain Müller, in which the students worked from natural sound samples collected in the city, and digitalized and processed by computer. They created in particular, the sound track for the video film "Utopia — the city".

However, the areas that for the past few years have occupied the most space in the EPI project are robotics and telematics.

3.4 Robotics and Telematics

Within the framework of educational robotics, every year since 1987, the Ecole Active de Malagnou has undertaken a global project.

A global project is a theme that all the teachers of a class agree to work on together throughout the school year. This is an important 'pivotal' theme, which may completely rearrange or even replace the normal school 'program'.

- 1987-88 *Handicap*, project concerning the quality of life of disabled people living in the city, experiencing living with a disability, meeting with disabled people, visits to places that have or have not been made accessible to people with some disabilities).
- 1988-89 *Utopia*, or how to dream up a city, urbanism, architecture... (group of three Genevan institutions).
- 1989-90 *Aqua*, study and modelization of the water cycle in Geneva.
- 1990-91 *Europa*, study of Europe, bringing together nine schools in five European countries (Denmark, France, Italy, Portugal, and Switzerland), resulting in a one month long exhibition at the Cité des Sciences de la Villette, in the outskirts of Paris.

- Since 1991 *Pangea*, an educational robotics project on the theme of environment, which brings together classes from every continent of the world through telematics.

4 The Role of Educational Robotics

Within the framework of educational robotics, students of 10 to 12 years old work on global projects, in other words, on themes that require observation, visits, interdisciplinary boundaries, and that result in the creation of programmed robotic models or simulations (LEGO-LOGO).

These models use sensors and react to their environment — they are therefore not only automatons, but rather they are truly robots.

Representations of places or objects observed, or synoptic scenes, these models are tools of simulation. They create an environment a bit comparable to an expert system environment, which helps the students to formulate their own expertise.

4.1 Validity of a Functional Interdisciplinary Approach

Even if it is only motivated by the desire to bring together teachers from various disciplines around a single project, or by the desire to emphasize coherence within a school program, an interdisciplinary approach is justified. But the effects produced by these global projects seem to me to provide an even stronger justification.

For example, when building a model, the students need, for technical reasons, to have a clear understanding of scale - it becomes absolutely necessary that math lessons address this subject, and for several weeks students work on fractions, scales and percentages with great interest, as they discover almost simultaneously their practical applications, which is much more interesting than just learning their theoretical lessons. They find themselves in a situation of functional interdisciplinarity - interdisciplinarity provoked by need.

4.2 Networks and Telematics

In each project undertaken by the Ecole Active over the past years there has been a desire to remain open to the outside world, to collaborate with other institutions.

One of the technologies available today to establish inter-school relations is telematics.

The establishing of exchanges with other classes is motivated by several desires:
- to lead other schools to modify their instructional approach (dissemination of active methods)
- to exchange expertise with other teachers, other students (to discover other ideas, other skills, leading to mutual enrichment)

- to provide young students with mastery of powerful communication technologies in an intercultural interest, to encourage better understanding of others and more respect of differences...

At first, these exchanges were made through 'Minitel' between schools in French-speaking Switzerland and in France. In new projects encompassing schools in other countries (Europa) or even other continents (Pangea), the necessary means vary depending on the tools available in each school (Networks such as Internet, simple communication software and modems, fax...).

5 EUROPA — An International Project

5.1 The Project and Its Content

Europa was an educational robotics project in which nine schools in five European countries participated: France, Denmark, Switzerland, Portugal and Italy. Each class had to choose a site nearby to visit and study.
- The site had to have a direct relationship to Europe.
- Through this site, a complete study was launched (history, geography, current politics...).
- Each class maintained correspondence with two other classes. (The students wrote in their mother tongue and the class receiving the correspondence found the means to translate the messages. The teachers corresponded in English.)

At the end of the project, the models were exhibited at the Villette Cité des Sciences et de l'Industrie in Paris for one month. The exhibition was visited by more than 50,000 people.

5.2 How LOGO is Introduced

In our school, the choice was made not to teach LOGO as a computer course in its own right, but rather to present it through the programming of their LEGO® models (to students from 9 to 12 years old).

After a brief presentation of the main LEGO® parts that can be connected to the computer (lights, motors, sensors, meters) the students are asked to develop a series of skills (to get 'certified'), to progressively acquire the (very elementary) notions that would allow them to program these parts.

Most of the more complex notions and skills are built when it becomes a real necessity to do so (such as "I want the screen to explain what the model is doing..." or "There has to be sounds and music at the same time..." or "That would be great if the model started up when the visitor pushed on a button..." or "It has to keep working and start over again all by itself when the program is finished...") or "Does anyone have any idea what I'll have to do to get that result?..." or "How are we going to program that in LOGO?").

5.3 Pedagogical Highlights

Our school chose to study CERN (Centre Européen de Recherches Nucléaires).

5.3.1 Interdisciplinarity

During this project, we decided to construct the school program around the needs of the project. The problems encountered in the LEGO-LOGO workshops were taken up in math lessons (percentages, proportions, scale...). The questions asked during the visits to CERN were discussed in history or geography lessons. The decorative elements of the models were made in art and shop classes (Functional interdisciplinarity).

5.3.2 Communication

The decision to have the students communicate in their native language (with the receiving class responsible for translation of the messages) seems to us to have been a good choice.

On the one hand, the children of each country were able to see the interest of learning a foreign language.

On the other hand, the content of the messages was not diminished by the need to communicate without having mastery of the language.

Several times non-French speaking students (in particular, Italian speaking ones) were able to show their classmates that although they had some difficulty in French, they mastered a different language, quite useful to the class at that moment.

5.3.3 Challenges

To represent the particle accelerators, the students chose to symbolize them by lights that moved around. Confronted with specific technical difficulties that were not resolved in class, we asked the children to find solutions by using the resources available in their environment. They therefore found responses that were quite varied. Their job was then to explain clearly to the rest of the class the solutions (sometimes proposed by adults, parents, older sisters or brothers...) and then to form small working groups to realize their projects. In the end, each accelerator was represented in a quite different way.

For several groups, this meant building several models before developing one that worked perfectly, and that would be sound enough to function throughout the exhibition.

6 Some Thoughts on the Model Building Process

When faced with the task of representing the studied object (in this case, CERN) in the form of a robot, the child is thrown into a situation that seems to me to be particularly rich pedagogically.

The student is in a constant movement between the object being studied (the global theme) and its representation (the model). The student is thus obliged to *synthesize* (simplify (but faithfully) the object enough to make it into a robot) and to *symbolize* it (produce a robot whose meaning can be easily recognized by anyone).

In this task characterized by experimental trial and error (to which the LEGO^R material and the LOGO language are well adapted), the student goes back and forth between the concrete and the abstract.

During our EUROPA project something happened that demonstrated to us the tangible value of the children's work.

After the Paris exhibition, but before the destruction of the model (to use the pieces for other projects) we asked CERN to exhibit the model on their premises. The people who 'explain' CERN to their numerous visitors from around the world quickly realized that the model built by the children was an excellent support to the explanations that they offered. The model is a representation of what non-experts understood following a serious study of the site. It is a symbolic representation comprehensible to everyone. And on top of it, the model has a poetic quality that touches our sensibilities.

After one month of exhibition, CERN decided to acquire the model to keep it indefinitely on exhibition in the 'Microcosm' surrounded by 'high tech' models representing the particle laboratory's research activities.

LEGO-LOGO educational robotics is a further innovation at the heart of our school. It gives meaning to interdisciplinarity and represents a concrete realization of active methods, often putting the child into situations requiring research.

7 Pangea

Pangea is the name given to emergent lands when they still formed a single continent.

It is also the name of this project, born out of previous EPI projects carried out by the Ecole Active de Malagnou in Geneva, and initiated by a few educators and researchers throughout the world at a LOGO colloquium in Fribourg (Switzerland) in 1990.

It is a project which, by means of communication and new technology, will erase the distances between continents.

- The project will have children between the ages of 9 and 12 (approximately) communicating on the theme of ecology.
- These children will be chosen from the four corners of the globe so that the planet is geographically and culturally represented (South America, North America, Africa, Asia, Europe, Australia).

The whole project will pivot on the theme of ecology and new technology (A contradiction in terms? I'm not so sure!).

Each class will pinpoint a problem in their area. Two approaches are possible:

- A study of a technological project which respects the environment (reforestation, fuel regulation, an ecologically efficient waste water treatment...) or

- A study of man's relationship to nature which is unfavorable to the environment (pollution, excessive deforestation, energy waste...).

Through the study of their chosen problem, the pupils will become familiar with all facets of ecology.

Throughout the project, each class will correspond with other classes. The students will construct a scale model representing the ecological problem they are studying. This can be an exact copy of a place or situation, a symbolic representation, or even a mechanical construction explaining the phenomenon. This model can be constructed with whatever is at hand using a maximum of leftover materials such as bits of string, plaster of Paris, aluminum, pieces of wood with cost next to nothing. Only the mechanical parts needed for automation and interface between the computer and the model will probably have to be bought: a few wheels, gears, small electric motors etc. The model will be driven by IBM compatible computers or Apple IIE or GS, which are relatively inexpensive, or can even be easily borrowed. Lastly, an inexpensive modem (Hayes 1200 or 2400 bauds) or perhaps minitel or fax machines belonging to a consulate or local cultural center, and the project can begin.

Once the place or topic has been studied, the scale model has been built, and once the correspondence is underway, the next step will be to have the children try to pilot each others' models by remote control through the use of a modem. In spite of the great distances between countries, this should be feasible for children from 9 to 12 years old.

The message?
- "At 9 years old, I can control computers and the technologies of our times."
- "At 9 years old, I can communicate instantly with someone at the other end of the world."
- "At 9 years old, I can make things move at the other end of the world."

Our 'Spaceship Earth' is very small. We have powerful means available to us. What do we, the children, want to do with them?

The Pangea project has three educational avenues.

Through their classwork and research, the students will be plunged into functional interdisciplinarity: geography, languages, science, mathematics, etc.

In the creation of their simulations, the students will learn robotics. First, through the conceptualization or modelization of the object of their research, they will be obliged to synthesize and symbolize their topic. Then, through the actual construction of the model, they will develop manual and technical skills, and an aptitude for creative trial-and-error experimentation. They will learn how to program with LOGO, and how to pilot another model over great distances.

And lastly, through correspondence, from traditional mail to electronic mail (Fax, Bitnet,...), from short wave radio to telecommunications, the students will communicate with children from across continents and oceans. Their correspondence will be general, getting to know each other and each other's customs; and specific, technical correspondence related directly to and LEGO® issues.

The teachers will corresponds as well, creating "training" networks to discuss pedagogical issues and resolve common problems, using English as a common language, if necessary.

Pangea will be coordinated from Geneva by the Ecole Active de Malagnou, with the help of national coordinators in eight countries: Argentina, Brazil, Costa Rica, Israel, New Caledonia, Spain, the United States and Uruguay.

Work has already begun in some classes. The project is expected to last at least two school years.

The Pangea project promotes interculturalism, communication, discovery of the 'other', a sense of power and ... responsibility.

I hope that this kind of project will be supported by people from the scientific community, philosophers, geneticians, researchers, ecologists ... and by important institutions who are sensitive to this approach, who are in a position to help provide the necessary means to carry this adventure out.

References

Bres, J.C., Environnement Pédagogique Informatisé, Cahiers pédagogiques de l'école active de Malagnou, 1987.

Bres, J.C., Europa, une expérience de robotique scolaire , Cahiers pédagogiques de l'école active de Malagnou, 1990.

Bres, J.C., La robotique comme ressource de construction de la pensée, communication au Third LOGO Congress, Petropolis/ Rio de Janeiro, Brazil, September 1992.

Bres, J.C., Dunand, P., Projet LEGO LOGO robotique, communication au colloque "Logo et Apprentissages" Université de Fribourg, 3-6 October 1990.

Doyle, M.P., Euro Logos, British Logo User Group, volume 1, 1992.

Not, L., Les pédagogies de la connaissance, Paris, Privat, Collection Sciences de l'homme, 1992.

Could the Robotics/Control Technology Be an Interdisciplinary Tool in School?

Montse Guitert

University of Barcelona, Facultad de Pedagogia, Dept. Didáctica y Organización Escolar c/ Baldiri Reixac, s/n Torre D 4°, 08028 Barcelona, Spain

Abstract. This text presents a robotic proposal. I have been in contact with a lot of experiences carried out in Barcelona concerning this field. We chose one of them to be followed in detail: the Pompeu Fabra school. The 7th level has been observed since we thought this was the best level to work with robotics.

The proposal is to set up an automatic classroom. For this we need one computer in the classroom linked to a BSP board which would be connected to several motors and sensors. All this equipment would be controlled by some LOGO procedures elaborated by students with the help of the teachers. Many tasks have to be carried out initially and a lot of very different disciplines are involved in this robotics work.

In fact, having a computer available in the classroom for anybody at any time could help to ensure that technology in school would not be only an isolated experience but be a new and powerful tool in the curriculum that could even favor a new philosophy of education or a pedagogic change.

Keywords. Cognitive tool, Computer language, Constructivism, Control technology, Cooperative learning, Elementary education, LOGO, Pupil learning, Pedagogical robotics.

1 Introduction

Robotics or control technology is presented as an element that involves a lot of aspects of the child's life, that would be part of his/her future and of a more and more automatized society. It is important in the curriculum area, not only as a working tool, but also as an object of real knowledge providing a methodological basis needed to improving the process of knowledge.

Robotics and control technology, being a study tool that brings together different knowledge settings, offers the possibility of cooperation with the different curriculum areas in an interdisciplinary way. It could be an important instrument in school to help the development of knowledge in the following way:

a) elaboration
b) invention
c) creation
d) imagination
e) analysing
f) researching
g) reasoning ...

2 The Starting Point:
What Had Been Done in the Field in Spanish Schools?

2.1 Infant School Robotics Workshop

2.1.1 Description of the activity

The computer workshop activity was born with the idea of introducing computers in the primary year students' timetable, with one class a week.

The pedagogic elements of the workshop introduce the child into a mythical technological atmosphere.

The LOGO language has allowed the creation of a microworld close to the children, with which they can draw whatever they want and do their projects.

2.1.2 Aims of the activity

- Demythification as a source of knowledge.
- Pedagogy based in the development of knowledge experience.
- School inserted into a social atmosphere in constant change.

2.1.3 Evaluation of the experience

The workshop has general principles: reflexion and logical solutions of the simulation and specific works on mathematics, language and plastics.

It has contributed basically to achieve the following aspects:
- A distance between the child and the object, aspect that is supposed to attend to a double space situation.
- The materialization of an abstract concept allowing for the difficulties that the fourth year students have with space concepts.
- Interrelation of an adequate technic in a global class context.
- The wealth of parallel works that the workshop has generated in other areas means that we can hope to create an interdisciplinary line with a lot of possibilities in secondary schools.

2.2 Computers in Primary Schools

2.2.1 Robotics workshop: Activity description

Robotic workshop has been thought of as an activity for 10 to 14 year old children, and is carried out in the school, directed by an external monitor.

Each of the seven workshops has a length of 10 days and it is done in two phases.

2.2.2 General aims

- To familiarize the students with robotics and its possibilities.
- To help the students to discover the importance of new technologies as a tool to facilitate work and human progress.
- To teach the students to value the advances that have been made in technology, but always with a critical point of view.

2.2.3 Specific aims

- Approximation to the technological manipulation of the setting as an interaction between the child and the physical world.
- To develop strategic projects of problem-solving from the manipulation of the robot.
- Demythification of the computer and the robot as omnipotent machines.
- Project design construction.

2.2.4 The final results

Each workshop will provide:
   a) A construction pattern model for children
   b) A group of procedures to the control of patterns made with LOGO
   c) An illustrated report of the workshop.
   The most important problem encountered was to obtain that the experience could be developed within the school framework.

2.3 The Flea Experience: Robotics Space

2.3.1 Activity description

The travelling workshop that was offered to the school was directed on students of primary and lower secondary schools, with a maximum number of 35 people per group. The activity duration was 2h 30min.
   The goal was to make an industrial setting as big as possible. The aim was that the students approach the differents tasks derived from an industrial process.

2.3.2 General aims

- To come into contact with these instruments.
- Make children realize that there is no such thing as a general purpose robot, but just designed robots to act on very limited settings
- To have children learn the computers-computing couple.
- To recognize that the robots are not doing anything by themselves, it is necessary to have a human "teacher" to introduce facts following a concise and concrete process.

2.3.3 Development of experiences

The experience begins with a careful observation of the setting. It explains the steps of the different positions and movements of the robot arm.

After that, there is a time for the student to try to assimilate all the movements with the motors and their commands.

Then, each level has a recommended task.

This activity was only a short demonstration of the robotic possibilities so far from school activities and reality.

2.4 Robotics Experience in a Primary School

2.4.1 Activity description

This is the activity in which I collaborated. It was conceived as a background workshop for the 12 years old students level, in the subject of plastic art.

It tries to support the continuity of pretechnology activities carried out in previous school years by the students: constructions, models of cranes, cogs and wheels etc.

It intends to help practice the electronics and computers (LOGO) knowledges studied in previous years.

The activity is developed in 10 sessions of 2 hours, each during a period of 3 months. Each workshop-group has 15 students and the class is separated into two groups.

2.4.2 Aims of the experience

- To understand the relation between speed, strength and energy in movement transmission.
- To learn about the different sensors and the instructions that control them.
- To recognize the different parts of the interface box.

2.4.3 Analysis and evaluation

The quantity and the quality of the concepts and abilities that the students have used in working with this activity deserve to be emphasized, whereas this activity was done in the subject of plastic arts with only this teacher subject, and whereas it was a short term experience.

3 Automatic Classroom

On the basis of my analysis and evaluation of experiences explained previously and considering how they respond to a short term situation, I will try to make a coherent proposal. I start with the aim of making the robotic and the control technology as an interdisciplinary tool. I understand the classroom as a reality, not only as a physical place.

3.1 Introduction

The proposal will be to make an automatic classroom, in which we have a computer in the classroom with an interface BSP plugged into the sensors and motors, that will control the LEGO® Technic or Fischer-Technik® buildings blocks designed for the students (to be helped by the teacher).

3.2 The Use of Sensors

To make an automatic classroom it is necessary to work with various kinds of sensors that will understand the information entry and will feed procedure for the computer with software programmed by the students.

This information will be analysed and used later in the context of different matters.

3.2.1 A light-sensor

When it does not detect any light, the computer switches on the bulb, according to the programmed software (Experimental Science subjects).

3.2.2 Temperature-sensors

They are used to:
a) measure outside temperature
b) control the inside temperature in order to act as a thermostat
c) control Fischer-house temperature
The framework is in the Experimental Science subject.

3.2.3 Two counter sensors

They count the number of the students of a class. The sensor detects each person that goes into the classroom and the program adds a unit in the counter and vice-versa: an other sensor detects each person that leaves the classroom and its program subtracts a unit in the counter (mathematics subject).

3.2.4 The primitive "time length"

When connected to a switch, it makes the bell ring each hour. This could be connected to some instrument that 5 minutes later closes the door or begins to count people again (mathematics subject).

3.2.5 A humidity sensor

It is used to control the plant humidity. When this is at a low level it tells you when you have to water them (Experimental Science subject).

3.2.6 An automatised black-board eraser

This could be a more difficult experience. We are working with different kinds of motors and sensors connected to the computer, that would begin to act by pressing a switch.

This is the first part of the generic purpose, and more at a technical control level. Now we propose it in a second part.

3.2.7 The construction of a mechanical instrument: a robot

The control technology will be used with the corresponding sensors and different motors, to use it as an instrument at school which the students of lower secondary school construct. This machine will have later a pedagogic use for the primary school students.

3.3 General Aims

- To understand robotics and control technology as a tool not as a goal, as a tool which is applicable to different settings.
- To learn to make new materials, instruments and pieces, developing educational experiences in an interdisciplinary setting in lower secondary school.
- To learn about robotics and control technology from its main characteristics: information, electronics and mechanics.
- To promote a methodological way of heuristic work, to create interest in the new technologies with the manipulation, discovery, invention, of how to make futures models.

3.4 Who Gives the Experience?

The team of teachers of the lower secondary level, with the specialist coordinator (physics/mathematics teacher, coordinator).

3.5 What Pedagogic or Educational Meaning Could it Have?

- To create parallel or similar situations to those in which they live in reality.
- To make contact with and to control different aspects in their environments that they might forget like temperature, humidity, light intensity, time ...
- To emphasize the initial aim of work with robotics as an instrument that helps the interdisciplinary aspect. The fact of having a computer in a classroom all day could make it possible not only for the students, but also teachers, to take part in the experience.

4 Conclusion

Firstly, this approach could lead to the use of computing in schools; not only as a short-term experience, but also in the future it could act as an educational philosophy and as a pedagogic change.

Secondly, robotics could act as a stimulant for some subjects which are more closely related, such as mathematics, physics, chemics, and also for other, more indirectly related, such as english for example.

Finally, we could say that robotics might go further than this and become an application for other subjects.

Let us remember that this is only a proposal. We need to know what the answer is in the reality.

References

Abelson, H., Disessa, A. (1986) Geometria de la tortuga. Anaya. Madrid
Aguareles, M.A. (1988) L'educacio davant la informatica PPU, Barcelona
Angulo, J.M., Curso de robotica Paraninfo.
Angulo, J.M. and NO, J. (1986), Robotica Paraninfo, Madrid.
Asimov, I. (1988) Jo robot, Trad, Ibarz, A i Marti, J. Pleniluni, S.A. Barcelona: Alella.
Bertran, Lawson y Jover, Hacia la comprension de la informatica Marcombo.
Billingsley, J. Robotica y sensors para el Commodore, Gustavo Gili
Bishop, B. Interfaces para el Spectrum. McGraw-Hill
CESTA Centre d'Estudes des Systèmes et des Technologies. Avancees. Robotique pedagogique.
Coiffet, Ph. (1981) Les robots. Tome 1. Modelisation et commande. Hermes
Coromines, J. (1976) Introduccion al control de procesos por ordenador. Universidad Gallego, Lowy, Mansilla, Robles (1985) Logo. SM Madrid
Garcia Ramos, Ruiz, F. (1985) Informatica y educacion. Informatica, Barcelona
Gros, B. (1987) Aprender mediante el ordenador. PPU, Barcelona
Marques, P., Sancho, J. (1987) Como introducir y utilizar el ordenador en el aula. CEAC, Barcelona
Papert, S. (1981) Desaffo a la mente. Galapago, Buenos Aires
Deval, J. (1986) Ninos y maquinas. Alianza, Madrid
Politécnica de Barcelona. Marcombo.
Ferrate, G. Robotica industrial Paraninfo.
Harrison, D. (1986) Robots, su estudio con ayuda del ordenador. Anaya. Barcelona
Hawkes, N. (1985) Robots y ordenadores. Marcombo
Laborda, J. (1986) Informatica y educacion. Laia. Barcelona
Logsdom, T. (1985) Robots: Una revolucion. Microtextos. Madrid
Lopez, P Y Neuma Foulc, J. (1987) Introduccion a la robotica II. Communicacion hombre maquina, programacion y control transformaciones homogéneas. Coleccion electronica y automatica. Arcadia. S.A.
Minsky, M. et al. (1986) Robotica. La ultima frontera de la alta tecnologia. Coleccion la sociedad economica. Planeta. Barcelona
Ogata, K. Ingenieria de control moderna. Prentice-Hall
Pawson, R. (1986) El libro del robot. Gustavo Gili S.A. Barcelona

Penfold, R.A. Técnicas y proyectos de interface. Anaya

Pentiraro, E. El ordenador en el aula. Anaya

Potter, T., Guild, I., Robotica. Anaya

Potter, T., Robots controlados por ordenador, Anaya

Reggini, H. (1982) Alas para la mente. Galapago, Buenos Aires

Rosello (1985) Logo, de la tortuga a la inteligencia artificial. Vector Ediciones, Madrid

Schmitt, N.M. / Farwell, R.F. (1988) A fondo: Robotica y sistemas automaticos. Anaya Multimedia

Sancho, J. (1986) La informatica en el curriculum. I Encuentro de Informatica en las aulas de Aragon, Navarra y Rioja, Utrillas

Sancho, J. (1988) Ordenadores, desarrollo cognitivo e innovacion educativa: La construccion de una falacia. ICE Universidad Autonoma, Madrid

Straker, A. (1986) Un deplorable estado de los hechos Times Educational Supplement, 9 May. Traduccion de Sancho, J.

Wat (1983) Aprendiendo con Logo. Byte Books

Actes del IV Seminari de Logo (1987), Associacio Logo. Universitat Autonoma de Barcelona

Logo, Metodologia y recursos educativos, (1987) MEC-Atenea, Madrid

Control Technology and the Creative Thinking Process for Teachers and Students

Marilyn Schaffer

Director, International Center for Education and Technology,
University of Hartford, West Hartford, CT 06117 USA,
Fax (203) 242-7002; Tel. (203) 243-4277
Internet: mschaffer@uhavax.hartford.edu, Bitnet: mschaffe@hartford

Abstract. To support teachers as they develop new skills and strategies in elementary science and technology we have established the International Center for Education and Technology. Our teacher development programs emphasize the use of LEGO/LOGO to help teachers use technology as a prime curriculum material rather than use computers and related technology as a dispenser of curriculum. We try to work with teachers to build from concrete activities to abstract thinking sequences in elementary science, involving the processes of analysis, reflectivity, and hierarchical conceptualization.

Keywords. Computer language, Constructivism, Control technology, Elementary education, LOGO, Longitudinal study, Project driven learning, Pupil learning, Pedagogical robotics, Teacher eduation.

1 Introduction

In many countries including the USA the time is ripe for *educational change*. *Teachers* know they must help students cope with the new challenges of our increasingly complex society, but many *feel overburdened and isolated* as they attempt the task. They are being asked to produce and deliver more and more, while many students are falling further behind in test scores and basic knowledge.

2 International Center for Education and Technology

To support teachers as they strive to develop new skills and strategies to help students prepare for life in the 21st century, we have established the International Center for Education and Technology at the University of Hartford. This Center has been established for the purpose of conducting programs in which *educational technologies* are used as a means *to advance and enhance learning opportunities for students and teachers world-wide*, especially in the areas of science, technology and intercultural understanding.

The *main activities* of the Center consist of providing *staff development programs* for teachers, administrators and teacher educators of our own and other countries to promote global awareness and communication in schools, and at the same time, provide enriched science, mathematics and technology education.

Our goal is to empower teachers to develop both the confidence and the competence to collaborate with us in the development of investigative project-based science curricula that can contribute to the improvement of science and technology education. With our *team of scientists* from both within and outside of the University, we are helping teachers develop a comfort with science. We try to foster the understanding that science can be complex—that interactions among phenomena are often deep and convoluted, that science phenomena may appear mysterious and unknown - but that it is these very traits that can make science so engaging.

We try to focus on the global implications of science issues. We emphasize the use of technology as a tool for the active "doing" of science and for engaging in collaborative projects in which teachers and students work together to plan, discuss and share their work with peers.

We view the technology that we use such as LOGO, LEGO/LOGO, and sensors as key ingredients in hooking both teachers and students into a subject that, at least in our inner city elementary schools, has not held much magic in the past. We have seen that the computers and related technology can serve as magnets to draw teachers to workshops and projects that they would not be attracted to if they carried only the label of "science." Somehow "science and technology" seems new to them, more interesting, and, it is our opinion, *less* threatening - it may be that since technology *is* new, teachers can not be expected to be experts.

For students we think that technology heightens the interest, and perhaps even puts a greater value on the science activity. It may seem more important since it involves expensive and sophisticated pieces of equipment. In many schools technology may be the only materially valuable objects that are entrusted to students.

3 Representative Activities of the Center

3.1 Constructionist Learning Project

Last year we completed a 15 month program which was funded by the National Science Foundation under a subcontract with the Massachusetts Institute of Technology (MIT). The first phase of the project was a three-week workshop for elementary teachers which was held in August, 1990. Forty Hartford area teachers were recruited in teams of 4-8 from each of four schools targeted for our grant activities. These teachers were joined by 12 educators from Eastern Europe and Latin America.

The major pedagogical theme of the summer workshop was "A Constructionist Approach to doing Elementary Science." We wove this theme into the activities of the workshop, including emphasis in presentations by Seymour Papert, the creator of LOGO, Stephen Ocko, a key developer of LEGO/LOGO, Robert Tinker, Chief Scientist of TERC, and other guest speakers who visited during the workshop.

During the second phase of the project, which took place over the academic year of 1990-1991, the teachers were helped to organize into teams and to collaborate on designing projects which they introduced in their classrooms, with the on-going support of the educators and scientists on our staff. During the year we made periodic visits to the schools to work with the teachers and their students, and conducted a series of Saturday workshops to reinforce the science and mathematics content of the teachers' projects. Our workshops emphasized hands-on explorations of materials the teachers would be introducing to their students such as LEGO/LOGO and "Fast Plants."

3.2 Project T.T.E.C.

Currently we are engaged in a three year project, funded by the National Science Foundation, entitled Project TTEC - *(Teachers, Technology, Environmental Concerns: Formula for Real Science in the Elementary School)*. We are collaborating with elementary teachers, their students, and the administrators of three inner city Hartford elementary schools, to build a science culture within each of the schools to promote environmental literacy, and hopefully, uncover the joy of doing science.

The project is being conducted in three year-long cycles, each of which has two time frames of activity: *Academic Year (Construction and Revision)*, during which project staff work with teachers and their students in the schools, and *Summer (Expansion)* when teachers participate in seminars and workshops.

Our objectives for the project include: Using computers and related technology including telecommunications as a scaffolding to create a Center for Environmental Science in each school; *Utilizing LEGO/LOGO* as a hands-on way of approaching topics such as solar energy and environmental monitoring; *Establishing and maintaining telecommunications links* among the participating teachers and our project team to support curriculum development and to establish collaborative relationships; *Evaluating the impact and effectiveness of our approach to science,* our teacher development methods, and our support activities, including our telecommunications thrust and our use of control technology; and *disseminating our results*

4 Control Technology and Teacher Development in Science

In our teacher development programs we use LEGO/LOGO to help teachers use *technology* as a prime curriculum material *in itself* rather than viewing it as a *dispenser* of curriculum. Through this material our teachers have begun to build a history of experiencing, and thereby, understanding, the *constructivist approach* and how it applies to science learning.

As teachers set out to create a mechanism that will be powered by solar energy or to make a robotic model of a sea creature, *they themselves pass through the steps* of exploration and creative problem solving that *we are asking them to achieve with their students.*

Many of the elementary teachers with whom we work are middle aged, female, and do not view themselves as mechanically or scientifically oriented. Working with LEGO/LOGO, they enter a process in which they focus, at a concrete level, on topics such as *gear ratios, motors, signals, feedback and control mechanisms, and problem analysis*, which normally they would not touch.

Working with LEGO/LOGO, *they not only touch* these things, *they succeed* in completing tasks which they have set for themselves, tasks which are complex and take them along uncharted routes, and which bring them enormous feelings of accomplishment. In essence *they have first hand experience in what we want their students to feel and think about school, about science, and about themselves.*

5 Control Technology and Student Progress in Science

Others with more experience than we have in working with students have documented the value of LEGO/LOGO in stimulating students' thinking about and understanding of control technology. We would like to corroborate these observations and add the following comments. We have found that *as students work* with these materials *they tend to generate questions that emanate directly from their interactions with the materials;* questions which might not occur to them otherwise.

For example, when a group of fourth graders were trying to program a mechanism that they had built, they asked, *"how does the motor really know when to turn on and off? It can not read the program, so how does it understand what to do?"* These questions led to explorations of the workings of switches—in their classroom, in their homes, discussions and demonstrations around the topic of electricity, and a series of experiments with batteries. The students *were pulled* into these topics *by their curiosities,* by *their interest to find out.*

We are trying to document the kinds of questions that students raise as they work with LEGO/LOGO and as they attempt to understand the very mechanisms that they are creating. Over the next year we are asking the teachers with whom we are working to help us collect material and to work with us to develop activities that can support students as they go about trying to answer the questions they generate.

2. Experiments and Case Studies

2.2 Methodology

Integrating the Use of LEGO-LOGO into the Curriculum of a Primary School: A Case Study

Maria Lucia Giovannini

Dipartimento di Scienze dell'Educazione, Università degli Studi di Bologna, via Zamboni, 34, 40126 Bologna, Italy

Abstract. This paper describes a control technology project carried out over a four year period. In particular it offers reflections on LEGO-LOGO cross-curricula use in primary schools. The first year was dedicated to the training of a group of teachers. The following three years of the project were conducted with a class in a primary school near Bologna (Italy). The organization and methodology used in this longitudinal study aimed to encourage the active participation and cooperation of the children and to develop important cognitive strategies and skills related to multiple subject areas.

Keywords. Assessment, Constructivism, Control technology, Elementary education, Evaluation, Learning environment, LOGO, Longitudinal study, Problem solving, Project driven learning, Pupil learning, Pedagogical robotics, Teacher education.

1 General Framework and Aims of the Project

The theoretical starting-points of the setting-up and carrying out of this work are numerous, but the general theoretical framework of this research project is one of social constructivism.

Papert and his team follow the constructivist approach. Ocko states explicitly that, *"LEGO-LOGO is perhaps the best example so far of a constructionist approach to education. Both LEGO and LOGO are construction sets: LEGO using plastic blocks, LOGO using abstract blocks. In each system the set of primitives is carefully chosen to give children a great deal of power and flexibility as they explore and create. And each system appeals to the child's natural sense of play, creating an environment that associates pleasure with learning. LEGO-LOGO can serve as a bridge between the physical world of the child's experience and the abstract world of the computer."* (Ocko and Resnick 1986a)

In addition to the familiar LEGO®building blocks the LEGO-LOGO system includes an assortment of gears, wheels, motors, lights and sensors. The computer communicates with the LEGO® devices through a custom-designed interface box, which is connected to a parallel card in the computer. Information flows through the interface box in both directions. By entering LOGO commands and procedures from the keyboard it is possible to send information to the LEGO® motors and lights. It is also possible to receive information from the LEGO®touch and light sensors plugged into the interface box. The programming language used in LEGO-LOGO is an expanded version of LOGO. In addition to the traditional LOGO primitives and control structures (such as IF, REPEAT etc.), there are twenty or so new primitives: output primitives (such as ON and OFF), input primitives (such as SENSOR?) , and one new control-structure primitive (WAITUNTIL).

In view of the characteristics of this system it seems to me important to underline its potential as regards the learning process. I use the term *potential* to underline the fact that it is not the material in itself that provokes learning, but its appropiate use.

a) The opportunity of using a computer language related to actual needs or a particular problem illustrates the difference between learning a programming language with a precise motivation and learning one without a precise motivation. The difference is also relevant in respect to the use of LEGO-LOGO primitives compared with batteries to power the objects built.

b) The expanded version of LOGO, used in LEGO-LOGO, offers not only the possibility of piloting the models but also of producing with ease texts and drawings related to the activity. The children can, for example, draw the model on the computer. They can document their work as they go, making notes on useful facts and information, the difficulties encountered, hypotheses on how it works and the results.

c) An important factor is the type of emotional relationship associated with the LEGO® bricks. In fact, for the large majority of Italian children, they are a playing experience. At pre-school and at home they enjoy themselves constructing and dismantling objects of various types, continually inventing new objects as a direct result of the characteristics and possibilities the material offers. One can therefore assume that the LEGO-LOGO activities are perceived as a game (Ocko 1987) and that the connection between the acquistion of knowledge and the child's past experience is a positive one.

d) The manipulation of and experimenting with projects created by the children permit a close involvement in scientific methods. The important aspects of this method are the following: making explicit the hypothesis and the decision making strategies; the anticipation of the effects of planned actions; the verification of these hypotheses through direct observation; thinking about the process used and about the effects produced; and learning from errors made. The possibility of starting with a manual construction exercise and of proceeding in the direction of higher levels of abstraction and codification allows the natural process of learning to be realised.

e) The production of models similar to real machines can stimulate thinking on their working and use. The inventors of LEGO-LOGO confirm that such activities can provide a more meaningful and motivating context for learning the traditional science curriculum and introduce elementary school students to important engineering and design concepts that are rarely addressed in today's curricula. (Ocko, Papert and Resnick 1988a)

There is also the possibility of relating the models constructed by the children to other economic, historic and social contexts in which they are used, adopting a cross curricula approach.

f) The need to plan, to choose the pieces, to build, to connect the models to the interface, to write computer programs to control the models, to test them and to carry out modifications, all create the need to compare and collaborate with others.

The inventors of LEGO-LOGO have described the results of the introduction of LEGO-LOGO at the Hennigan School in enthusiastic terms. They have fully verified numerous assumptions and they have underlined how in a similar environment the children classified as *learning disabled* changed their behaviour and showed marked improvement in other subjects (Ocko 1987).

We intended to explore the realization of these possibilities within the scholastic context of the Italian primary school. A context in which the scolastic robot has not been introduced and where activities tend to be more structured than in other countries. We thought the approach to science should more actively involve the children in experiments and in hands-on activities.

The first phase of the research was carried out in 1988. It was funded by the EEC and conducted in collaboration with the University of Liège (Belgium) in association with Ocko (M.I.T.).

The aims of the research were as follows:
1) To set up a robot environment in a primary school.
2) To study the reactions of the teachers and the children to the LEGO-LOGO environment.
3) To offer pedagogical and technical help in setting up a LEGO-LOGO environment.

In Bologna teacher training was the object of the first phase of the research before conducting the activity with the pupils.

The subject of this paper is related to the second phase of the research. During the three years during which the study was carried out in the classroom we intended to explore the use of LEGO-LOGO in a primary school and reflect upon the process of teaching/learning in this educational environment. The study of the link between the LEGO-LOGO environment and primary school curricula was one of the aspects of the research project.

2 Characteristics of the Classroom and School Environment

The exploratory classroom study began in the 1988/89 school year, with 17 children (8 to 9 years old) in their third year of elementary school. The project ended in June 1991. The school in which the experiment took place is in Riale, a town of a few thousand inhabitants on the outskirts of Bologna. In the school there were five classes with a total of 71 children in 1988/89. The school, with all its full time classes, is known for its openness in respect to innovations. Within the school there are numerous laboratories for various types of classes, for example printing, woodwork and theatre. Of particular interest is the mathematics laboratory for problem solving and the computing laboratory. A student in education and myself, the coordinator, also took part in the LEGO-LOGO activities. At one of the last meetings an electronics engineer with a knowledge of LEGO-LOGO participated with the purpose of analysing thoroughly some of the children's questions and studying with them the models they had produced.

3 A Collectively Developed Itinerary

The insertion of the LEGO-LOGO activities into the basic curriculum was one of the aims of the first phase of the research, but it was studied only in general terms. Together with the teachers involved in the research we started to construct a work plan adapted to the context in which we were to carry out the project and that would utilize the characteristics and possibilities that LEGO-LOGO offers.

The aspects of the curriculum to be taken into consideration did not only consist in the development of concepts like mechanics but also of other subject areas such as linguistics, social studies and economics in addition to the development of important skills and cognitive strategies. Much importance was also attached to the development of collaboration skills and attitudes. Among the multiple interactions taken into consideration were the children's own attitudes to words' stereotypical gender behaviour.

The role of the teachers in organizing the activities with the researchers was not only relevant in the initial phase, but also for the duration of the project. The meetings with the children were followed by formative evaluation sessions with the teachers and the researchers. They analysed the information gathered, discussed their own and the children's conduct and planned the following LEGO-LOGO session. This plan was written up on a schedule that was compared at each meeting with the schedule of the completed session. The frequency of the meetings between the teachers and researchers was decided depending on the particular phase in progress, the kind of problems encountered during the classroom activity and the (possible) necessity of using other instruments and strategies for formative evaluation. The meetings were characterized by a high level of collaboration considering that one of the subjects discussed was the conduct of each of us in relation to the children. The criteria of the analysis of these relationships was based on the ability to encourage the individual child to play an active role in the learning process and to stimulate a collaborative working environment both within the subgroups and the class as a whole.

4 The Activities Carried Out

4.1 Becoming friendly with the material

After an initial phase of playing with the LEGO® bricks during play periods the class was divided into two groups. They worked simultaneously. One group was followed by the mathematics teacher and the other by the computing teacher. The teachers presented the children with the LEGO-LOGO components including the interface box, its operating instructions, the cables and the interface card plus instructions on how to put them together. They demonstrated two example models, a car and some traffic lights, which they had built themselves. Each teacher coordinated one of the two groups into which the class had been divided. In this phase the teachers acted as catalysts helping the students to set the materials up, encouraging them to describe and reflect upon the effects of the programming commands, the function of each LEGO® piece and to discuss and compare, using empiric checks, the different ways the models worked. To stimulate thinking on the functioning of each piece of the model a report card was used, which could be filled in individually or in pairs. Then it was discussed with the whole class. The successive sessions demonstrated the importance of this presentation. Initial discussion provided the children with the possibility of creating and exploring on the basis of hypotheses and not simply proceeding by trail and error.

4.2 The first year projects

After these sessions the children were free to create and build a project in agreement with the other members of one of the four sub-groups. The number of subgroups was determined by the quantity of sets of LEGO® Technic building materials available even if we would have preferred to have smaller groups.

At the end of the first year the five projects realised (one subgroup built two) were three child robots, a jeep and a Spitfire. Each one with one or two motors, flashing lights and a sensor. The focus however was not on the completed project but on the process of producing it.

During the building of the models the teachers tried within the subgroups to stimulate the children to explain their choices and hypotheses, to construct general principles and specific knowledge as they worked and to test the efficiency of their design. As important notions and concepts arose, the teachers, as discussion leaders, provided structure and explanations for the whole class. Some topics for discussion were for example the working of the pulley, the gear, the rack, rotary and linear motion, friction, inertia, force, energy, speed, the measuring of time and so on.

4.3 The personal notebook

Each child built up over the three year period a personal *notebook*. It was used to sketch designs of their project, to keep a record of problems encounted, of various hypotheses, of solutions, of results obtained and of the *bugs* encountered while they were building and programming. In regard to the overcoming of problems the *notebook* was of particular help. For the most part the children used the computer to create this manual. Thus they used the computer in different ways and not only to control the models.

One of the aims of keeping the notebook was to stimulate not only intragroup but also intergroups discussion and communication. The use of these written reports to discuss and share ideas with the members of the other subgroups on the methods adopted and problems faced made it necessary that the reports be comprehensible to those who had not taken part in the construction of the model. The teachers utilized this opportunity to improve written skills within the class.

4.4 Longdistance communication in foreign language

On the basis of the children's need to correspond via EMAIL with a class of american children, the language teacher taught a unit on how to write a letter so as to enable the children to describe their projects and the procedures envolved.

Discussions, in which the whole class took part, on the problems encountered and solutions adopted allowed both a better understanding and the acquistion of oral expression skills.

4.5 From the subgroup project to a class project

While in the first year each subgroup was free to choose which model to build, in the second and third years a class project was developed in which the models built by the various subgroups were connected. The purpose was to overcome the excessive competition between the different subgroups.

The project chosen by the class was a building site. The children continued to work predominately in subgroups constructing a crane, a tipping lorry, a concrete mixer, a bulldozer and a fork lift. One of the planned activities was the building of a three dimensional model of a building site with hills made of isometric compensate in which to put the completed LEGO models. It would have been interesting to finish the site with other elements to make it more realistic but there was not enough time.

In the construction of the models other materials were used in addition to the LEGO® bricks so as to render the objects as similar as possible to the real thing. For example, here is an extract of one of the *notebooks* of the concrete mixer's subgroup: *"To make the barrel of the concrete mixer we used a stripe of plastic and the lid of a yogurt jar attaching them with staples to obtain something that looked like a concrete mixer's barrel. To build the ring nut we used a length of play-dough. We passed a gear over it to obtain the teeth which we redefined with a toothpick. To build the frame we took a thread of iron, we wrapped it around a*

screw in the foot of the frame, then we wrapped it around the axle behind the
barrel and lastly around a screw in the other foot of the frame."

4.6 The relations with real life and science curriculum

The children's choice of constructing a building site provided the opportunity of
preparing a visit to such a working situation. However it was not possible to
actualize the visit for safety reasons. It did not though stop the children from
documenting with photographs they took themselves showing the essential
workings of the machinery, the dangers of injury and the conditions of work. In
addition the researchers made a film providing the opportunity to observe at close
quarters a building site and analyse the machinery and labourers at work. During
the various phases of the activity, the children consulted these resources.

To complete the documentation that the children wrote with reference to their
work and their methods of work, photographs of the models were taken. The
children were also given schedules to insert into their *notebooks* which consisted
of drawings of the workings and constraints of the procedures. The documentation
helped the children to reflect on their methods of working, to think about real
problem solving situations and to overcome the difficulites encountered.

The structuring of important mechanical and technical concepts was facilitated
through lessons in university physics laboratories available to elementary school
children. In these sessions the children were very active in that they continually
asked questions based on their experiences. They were able to deepen their
understanding of weight lifting systems, levers, power and sloping planes found in
the LEGO-LOGO activities.

4.7 A constant gathering of data

As has already been said the setting up of and the type of activity chosen were
decided by the teachers and the researchers, based on information gathered on the
running of the sessions. This constant gathering of data did not only consist in
revealing the presence or absence of specific facts and behaviours but also of
understanding the context within which it took place. The methods and the
procedures of this formative evaluation were multiple. Besides the drawings and
the essays produced by the children, there were the photos of their models, the
recordings of conversations in the small and large groups, observation protocols
and observation grips, rating scales, interviews, audiotapes, videotapes of key
activities and questionnaires. The use of these multiple methods allowed the use of
triangulation as a strategy of interpreting the phenomena in order to overcome the
biases and the inherent weaknesses of single measurement instruments.

5 Main Results and Prospects

The skills of hypothesizing and predicting, of exploring ideas in depth, of reflecting upon the meaning of different *bugs*, of using errors as a source of new ideas, of working well with classmates, of maintaining a positive attitude when things did not work immediately are some of the indicators of student performance taken into consideration even at the level of final evaluation. These abilities were observed not only in the recordings of the LEGO-LOGO activities but also by the teachers in other types of activities.

At the end of the first year, in order to measure skills acquired by each child, particularly in relation to the functioning of the interface box, the computer and scientific concepts encounted during the LEGO-LOGO activity, several tests were administered. The results showed that the questions were answered with ease by the children.

When the models were completed each member of each subgroup made a presentation to the rest of the class explaining the workings of the model, the procedures used, the problems encounted and the solutions adopted.

In the final sessions more tests were administered, some of which were also given to a control group, to verify the acquisition of skills such as the use of hypothetical verb tenses, the description of and resolving of problematic situations not related to LEGO-LOGO and also logical and scientific concepts.

In the case of scientific proof, the multiple-choice questions were concerned with the effects of power, leverage, sloping planes, wheels, pulleys, gears, parallel and non parallel rotation axes and the transmission of movement.

The problematic situations were of two types, open and closed. The open one consisted of an illustrated description of a child lost in the woods who has to find his way home using the information provided. A written answer is required. The closed one is a science fiction story about the pilot of a spaceship who has to solve a series of problems. The questions were multiple-choice that did not require specific technical knowledge but an understanding of the correct approach to problem solving.

The results of these tests, even if for some items the percentage of mistakes was high, were better than those of a control group in which LEGO-LOGO activities were not used.

The teachers' and children's enthusiasm for the activity was maintained at a very high level for the duration of the research. This was reflected by the teachers' contribution by the children's behaviour and by the answers to the frequently administered rating scales. The experience allowed the acquisition of knowledge and skills connected to various subjects areas by means of an enjoyable activity. In particular, concepts such as mechanics, which girls often consider to be distant and hostile. It also provided a *privileged* setting in which personal contributions could be valued, particularly those which appear to be less in keeping with tradition approaches to activities of a technical/information science nature, for example, design, proportion, aesthetics which are normally undervalued. The results of the final evaluation highlighted the initial overcoming of some misconceptions and widespread stereotypes.

Acknowledgements

The realization of this research project was largely made possible by the contributions of C.N.R 89.02200.08 and 90.03526.CT08.

References

Ackermann, E. (1987) Lego-Logo activities: A formative evaluation. Cambridge: M.I.T.

Adamson, E., Helgde, G. (1989) Exploring art and technology. In: G. Schyten, M. Valcke (eds.) Second European Logo Conference. Proceedings, Gent 1989, 414-427

Browning, C.A. (1991) Reflections on using LEGO tc Logo in an elementary classroom. Third European Logo Conference. Proceedings, Parma 1991, 173-185, Parma: A.S.I.

Christensen, C. (1985) Robots légo controlés par ordinateurs et interfaces. Communication à l'université d'été "Les N.T.I. et l'enseignement primaire". Liège july 1985

Denis, B. (1987) Technologie de controle et LOGO. La robotique, ses enjeux, ses modalités. Education-Tribune libre 208, 61-67.

Denis, B., Solot, F. (1986) Essai de définition et de classification de quelques ojectifs poursuivis dans l'environnement LOGO. Liège: Laboratoire de pédagogie expérimentale de l'Université

Giovannini, M.L. (1988) La scelta Logo: alcune riflessioni. Compuscuola 29, 86-89

Giovannini, M.L. (1990) Robotique pédagogique à l'école primaire. In: M.Vivet (ed.) Actes du premier congrès francophone de robotique pédagogique. Le Mans 1989. 17-28. Le Mans: Centre d'Impression et d'Edition de l'Université du Maine

Giovannini, M.L. (1991) Teachers' strategies and children's learning in a LEGO-Logo project. In: E.Calabrese (ed.) Third European Logo Conference. Proceedings, Parma 1991, 669-674, Parma: A.S.I.

Giovannini, M.L., Baroni, F. (1992) Bambine, computer e ingranaggi. Scelte e interrogativi relativi a un'esperienza in terza elementare. In: E.Beseghi, V.Telmon (eds.) Educazione al femminile: dalla parità alla differenza. 290-298, Firenze: La Nuova Italia

Ocko, S. (1987) Come integrare le construzioni LEGO con LOGO, Compuscuola, 19, 70-75.

Ocko, S., Papert, S. and Resnick, M.(1988a) A learning environment for design. Cambridge: M.I.T.

Ocko, S. Papert, S. and Resnick M. (1988b) Lego, Logo and science. Cambridge: M.I.T.

Ocko, S. and Resnick, M. (1986a) Integrating Lego with Logo: Making connections with computers and children. Cambridge: M.I.T.

Ocko, S. and Resnick, M.(1986b) Lego/Logo software. Cambridge: M.I.T.

Papert, S.(1984) Mindstorms. Bambini, computers e creatività. Milano: Emme ed.

Pea, R.D. and Kurland, D.M. (1983) On the cognitive prerequisites of learning computer programming. New York: Center for Children and Technology

Resnick, M. and Ocko, S. (1987) Lego/Logo and science education. Cambridge: M.I.T.

Resnick, M., Ocko, S., Papert, S.and Silverman B. (1988) LEGO/Logo. A learning Environment for design. Cambridge: M.I.T.

Solot, F. and Denis, B. (1987) Deux grilles d'observation de comportements pour augmenter l'efficacité des acteurs de l'environnement LOGO. Forum LOGO 2, 53-55.

An Aesthetic of Learning Environment Design

Gregory Gargarian

Institute of Technology Massachusetts, 20 Ames Street, 02139 Cambridge, Massachusetts, U.S.A.

Abstract. In this paper I discuss research on LOGO-based microworlds for constructing robots as a means of articulating *constructionist interactivity*, an aesthetic of learning environment design and evaluation. In this way self-regulation can be recursively linked to learning, the environment of learning and the development of environments of learning.

Keywords. Constructivism, Control technology, Design, Elementary education, Evaluation, Learning theory, LOGO, Microworld, Programming, Pupil learning, Pedagogical robotics.

1 The Mind as a Cybernetic System

Piaget's theory of intellectual development is a theory of mind as a cybernetic system with states of stable equilibrium separated by states of unstable equillibrium. This can be illustrated by Piaget's classic water pouring experiment in which he asks children whether the amount of water is the same or different when it is poured between two containers with different shapes (below). *Pre-conserving* children - i.e. those who answer "no" - will let their incomplete perceptions overwhelm the logic of the action by attending only to the height of the containers ("the second container is taller"..."it's shorter"). However, because these children are consistent about their answers, the knowledge that supports these answers is *stable*.

1.1 Conservation of Continuous Quantities

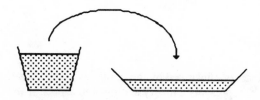

Fig. 1. Piagetian experiment no. 1

Is the water in each beaker the same? Why?

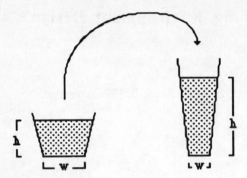

Fig. 2. Piagetian experiment no. 2

How about in this case?

Some children, still not conservers, can concentrate on different dimensions of the container but do not recognize the connection between them. While the water is being poured they are able to predict that the quantity of water will remain unchanged; yet, once they observe the results of the action these same children will change their answer. Because their answers are <u>not</u> consistent, the knowledge that supports these answers is *unstable*.

Only children with conservation of liquids consistently answer "yes". They understand that height and width compensate for each other. These children have made a permanent connection between previously disparate knowledge. Knowledge is once again stable, but at a different organizational level.

1.2 Equilibria

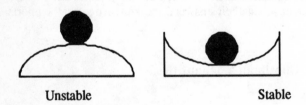

Unstable   Stable

Fig. 3. Equilibria of knowledge

Piaget's cybernetic model of mind can be illustrated by the drawings given above. In these drawings the ball (in black) represents a theory about some aspect of the world, and the foundation (beneath it) represents some organization of knowledge on which the theory rests. Both balls are in a state of equilibrium However, the balanced ball on the left is highly unstable; a little push will send the ball rolling. By analogy, because the theory rests on "slippery" foundations, it will immediately be put in question. The ball on the right is highly stable; most pushes will return it to the middle of its foundation. By analogy, because the theory rests on strong foundations, it will not be challenged by (alleged) counter-evidence.

2 Microworlds and Powerful Ideas

Piaget largely ignored the states of unstable equilibrium that separate his mental stages. His experiments were designed to identify the current state of a child's knowledge only. However, neo-Piagetians observed that a child's interaction with a Piagetian experiment could facilitate the acquisition of knowledge under examination by the clinician. From this new perspective, a Piagetian experiment could be seen as a "technology" for de-stabilizing thought strategically in order to promote its re-organization at a higher level (Ackermann, 1991).

One might even conjecture that Papert's notion of a microworld has its intuitive roots in such experiments. Like Piaget's experiments, *microworlds* are virtual worlds in which a learner is introduced to ideas with broad applicability (like conservation). Papert calls such ideas *powerful ideas*. Powerful ideas are made concrete because they are embodied in a medium - in Piaget's case an experiment, in Papert's case a computer-based learning environment.

A classic example of a microworld is Andrea diSessa's dynaturtle, a cousin of the LOGO turtle. A LOGO turtle is a graphic object that can be navigated around the computer screen using commands like FORWARD or BACK some number of steps, or LEFT or RIGHT some number of degrees, both specified by the user. A LOGO turtle is used in a turtle geometry microworld to explore powerful ideas in mathematics and computer science like variables, recursion, modularity, and abstraction.

2.1 LOGO's screen turtle (▲) executing commands that produce a square

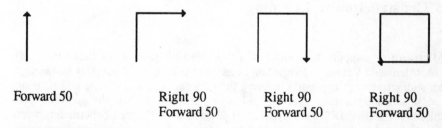

Forward 50	Right 90	Right 90	Right 90
	Forward 50	Forward 50	Forward 50

Fig. 4. Screen Turtle

A dynaturtle is a reprogrammed LOGO turtle that operates by the laws of Newtonian motion (a powerful idea). Instead of moving by forward steps it moves by KICKS (given to it from different directions and with different strengths). A dynaturtle conserves momentum. Kicking the dynaturtle has an effect similar to striking a billiard ball in motion with another one; i.e. the new KICK is "averaged into" the momentum of the one in motion.

2.2 Giving a "kick" to the billiard ball in motion

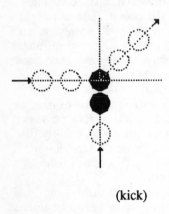

(kick)

Fig. 5. Dynaturtle

While our earth-bound world does not provide pure examples of Newtonian motion, dynaturtle's computational world does. When a child navigates a dynaturtle on the computer screen, he begins to have experiences with friction-free objects and can use these experiences to construct intuitions about Newtonian motion [see Note 1: Dynaturtle].

3 Constructionist Learning

Microworlds support new kinds of natural learning, or what Papertians call *constructionist learning*. Piaget taught us that learners construct their knowledge by interacting with the world, making theories about it, and testing and revising the theories until they discover what is invariant within the world. What microworlds (like DiSessa's dynaturtle) do, and what Piaget's experiments were not designed to do, is to provide learners new kinds of worlds (like Newtonian motion) with which to interact.

Microworlds are different from Piaget's experiments in an important way.

They provide learners the means to construct operationally reliable ("stable") thought because the actions required to operate within the microworld have links to the mental agencies that inform these actions. In other words, the most productive actions reflect the powerful ideas that one would need to have in order to take these actions (McGee).

Microworlds also depart from Piaget's stage theory. They assume a theory of learning in which development advances in an uneven rather than uniform way, that the mind is composed of many cybernetic systems. Piaget, himself, noticed this *decalage* of development, the fact that there is a time lag between the acquisition of different kinds of conservation (e.g. between substance, weight and volume). One might say that microworlds support micro- rather than macro-development.

Where Piaget was interested in how learners naturally develop, Papert has been interested in making new kinds of intellectual development natural. Papert's interventionist strategy can be illustrated using the ball-foundation analogy, now constructed into a pattern of cascading "micro-stages" (below). The arrow that pushes the ball to the left represents an intervention in thought provoked by the learners interaction with a microworld. This intervention hopes to provide the (cognitive) potential for a skip to a new micro-level of knowledge.

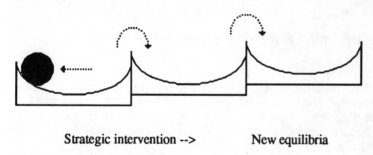

Strategic intervention -->   New equilibria

Fig. 6. Development of knowledge

4 Cybernetic Microworlds

In LOGO, learners can construct simple procedures and use them as subprocedures in more complex programs. In LEGO®, children can construct sub-assemblies and use them to build more complex structures. LOGO researchers recognized a kindred spirit in LEGO®. By constructing communications between LOGO and LEGO® motors, gears and sensors, researchers see opportunities for introducing children to cybernetic microworlds. Now the highly engineered floor turtle (of early LOGO work) is replaced by robots designed by the children themselves. Virtually any robot creature constructable in LEGO® can be navigated with LOGO programs.

Our intellectual partnership with LEGO® is adding new dimensions to questions about the relationship between models and reality. For example, the same LOGO commands for producing a square using the screen turtle (presented earlier) could produce the following "square" using a child-constructed robot.

Fig. 7. A robot executing a square procedure

Are the imperfections in the resulting drawing attributable to bugs in the square program? In one sense they are not. Clearly the screen turtle provides a faithful execution of the model of squareness captured in the LOGO program. But in another sense they are. Should not one's model of squareness incorporate the problems of friction encountered by the robot as it moves in the physical world?

5 Debugging and Objects-to-Think-With

Researchers have assimilated this question into a different intellectual framework nspired, in part, by robotics research (WOLKOMIR) and the fantasy machines described in Valentino Braitenberg's *Vehicles* (Braitenberg). Consider the following schematic drawing of a robot with a body, wheels controlled by motors (themselves, controlled by LOGO programs), and two photo cells.

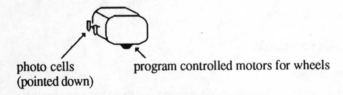

photo cells   program controlled motors for wheels
(pointed down)

Fig. 8. Self-regulating Robot (side view)

The photo cells are pointed downward so that they can be used to follow a square light strip (left and below). The width of the strip is narrow enough so that the two photo cells straddle it. If the robot veers to the left, the right photo cell will "notice" the strip and direct the robot to turn right a little. Conversely, if the robot veers right, the left photo cell will "notice" the strip and turn left a little. If we were to draw the path of the robot tracking this strip, we might see something like the drawing shown on the right (below).

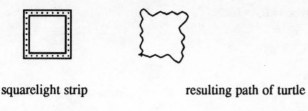

squarelight strip   resulting path of turtle

Fig. 9. A self-regulating floor turtle following a square light strip

Obviously, the precise notion of squareness has degraded from the screen turtle to the path of the self-regulating robot. However, the importance of squareness has also diminished. In the case of the turtle geometry microworld, the child *plays turtle* by walking in a square and *teaches the turtle* to draw a square by describing his path in LOGO steps and turns. The resulting square is then used as feedback for him to debug his square program. In the case of the cybernetic microworld, the child can also *play robot* and teach it how to modify its own responses to sensory cues. However, with the robot, the child debugs the debugging strategies of the robot!

Both microworlds have what Papert calls *object's-to-think-with* (a screen turtle in the first case, a robot in the second). By playing turtle or playing robot, the learner can project his own behavior onto these objects as a means of making thinking concrete. Equally important, both microworlds allow learners to regulate their own behavior through interactions with the world. In the case of the robotics microworld, it is a self-regulating learning environment to explore the powerful idea of self-regulation itself.

6 Synthetic Ethology

LOGO's electronic bricks add richness to the LEGO-LOGO world. They are wireless and there are many types of electronic bricks; action bricks (motors, lights), sensor bricks (light and sound sensors), logic bricks (and-gates, flip-flops, timers), and programmable bricks on which LOGO programs can be down-loaded (Resnick and Martin).

Consider the following robot constructed from electronic bricks. Like the previous one, the light information from the two photo-cell "eyes" control the motor behavior of the robot. However, with this robot, its "eyes" are pointing forward rather than downward and a light shield between the "eyes" causes incoming light (from a flashlight) to be registered differently on each "eye" depending on its incoming angle.

program controlled motors for wheels

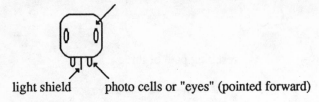

light shield   photo cells or "eyes" (pointed forward)

Fig. 10. Self-regulating Robot (view from above)

The robot responds to light in three ways. If the flashlight is directly in front of it, the robot moves forward; if the flashlight is to the left the robot moves left; and if the flashlight is to the right the robot moves right.

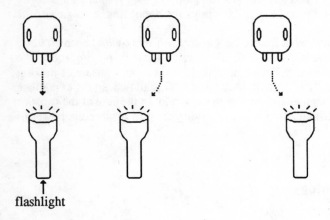

flashlight

Fig. 11. Robot's response to light (view from above)

Nira Granott, a researcher in our group, wanted to know how people think about robots like this one. She asked teachers to observe a number of robots ("weird creatures") and describe their behavior, much as an ethologist might do with real animals. She noticed that the teachers tended to use anthropomorphic descriptions first and only later began to give mechanistic descriptions. In part, her research has been about constructing a bridge between these two kinds of descriptions [see Note 2: Animistic Thinking].[1]

---

[1] To the above descriptions, Resnick and Martin add *informational* descriptions. For our light-sensing robot, an informational description might read: The robot's sensors quantify light inputs. These values are used as inputs to a program that sends a control signal to the motors directing the movement of the robot.

6.1 A Teacher's Anthropomorphic Description

The robot loves light.

6.2 A Mechanistic Description

If the left eye receives more light than the right, the robot moves left.
If the right eye receives more light than the left, the robot moves right.
If both eyes receive equal light, the robot moves forward.

For Piagetians, new learning is possible because one uses what is familiar to get at what is unfamiliar. However, if something is totally unfamiliar one cannot use a familiar mental scheme to identify unfamiliar data; and without data, one cannot construct a new mental scheme. This *learning paradox* expresses the fact that learning something completely new is impossible, or if learning is happening, what one is learning is not completely new (Gargarian). Granott's research shows how different kinds of discourse could be used to transcend the learning paradox.

Mechanistic descriptions are not necessarily privileged among experts. Anthropomorphic descriptions can capture the functionality of complex processes one already understands in mechanistic terms. For example, computer programmers often talk about modules of programs "wanting" particular kinds of inputs from each other. Programmers use these descriptions, not because they are unaware of the computational mechanisms that support them, but because they want to examine them at a level of abstraction where the modules interact. Knowing which kind of description will be most productive and when, is a kind of knowledge that one can learn to use to guide one's own learning.

7 Learning by Design

The best way to have experiences with geometry is to design geometries and discover their laws. This is what the microworld of turtle geometry supports. The best way to understand self-regulating systems is to design them and observe their behavior. This is what microworlds using LEGO-LOGO and the electronic bricks support. Even with dynaturtle where one does not produce an artefact, acting on the dynaturtle and observing its behavior is how one designs one's knowledge about Newtonian motion.

As a constructionist, I argue that the only way one learns is by designing! In my examination of the design process of composers I have learned that designing is not conventional problem solving. In computer science, problem solving requires well defined problems with recognizable solutions (like tic-tac-toe or chess). The problem is to develop effective ways to navigate within a "search space" of solutions one has already anticipated. In contrast, designing is exploratory, problems are ill-defined and solutions are often debatable (like car, chip, architecture or music design). Desiging is both the more elusive idea and the most pervasively practiced. The fact that some designers are better than others has caused me to ask what kinds of knowledge make the difference between novice and expert designers.

One answer is that experts are good at debugging. Writing is largely the art of editing; designing is largely the art of redesigning; and theory building is the art of theory revision.

Another answer is that good designers have considerable skill. I define *skill* as a combination of tools and cases. A tool can be either physical (e.g. hammers), conceptual (e.g. logic, formulas) or computational (e.g. algorithms). Case knowledge is the awareness of the kinds of situations in which a tool can be productively employed. Without links between a tool and case knowledge, there is no skill. Experts are better designers than novices because they have more kinds of tools and more kinds of previous cases from which to draw. Microworlds provide tools and opportunities for learners to acquire skill, case by case.

Finally, designers not only produce artefacts, they construct a design environment (or microworld) in which an artefact is designed. (Composers not only make a piece of music, they construct the composition environment in which they make a piece of music.) Because experts have learned to use a variety of tools and have extensive case knowledge, they customize a microworld of design so that the appropriate tools and case knowledge is readily available. We are only beginning to provide children with the facilities to design their own microworlds like expert designers do (Gargarian).

8 Tools for Knowledge Creation

Any tool--be it material, conceptual, or computational--that makes new kinds of interactions possible can facilitate the creation of new knowledge. In hindsight, we can see that science often advanced with the presence of new tools. The telescope, microscope, and cell staining techniques in neurobiology are a few examples of technologies that supported the creation of new thinking and practices by making the invisible visible or the abstract concrete.

The history of music provides other kinds of examples, important because they reveal some relations between tools and culture. The piano embodies knowledge about tuning systems and musical scale in its design. Its black and white keys provide a friendly interface for locating notes and octaves, and foot pedals for sustaining or muting strings; this gives performers the support they need to manage the piano's range of expressive possibilities.

The violin's shape, materials and varnishes combine to make a transducer superior to its cultural counterparts. Unlike the piano, violins can be easily retuned. Because there are no frets along its fingerboard the violin is quickly adopted by music cultures that use different tuning systems and scales.

Piagetians would say that the violin, more than the piano, was easy for other cultures to *assimilate* because the native music could be performed on it. Some anthropologists have said that the piano was a *westernizing* instrument because it introduced a new practice while the violin was a *modernizing* instrument because it amplified past practice (Nettl).

9 Supporting Tool Use with an *Operational Folklore*

New knowledge is increasingly becoming a process of developing tools and learning how and when to use them. Conventional notions of "technology transfer" (to schools, industry or countries) usually ignore the process and consequence of technology assimilation. Tools exist in some relation to culture. The technologies in themselves are not the new culture; rather, they provide support for the creation of new culture or the amplification of old culture.

LOGO researchers have learned to appreciate that the culture around the use of a technology is as important as the technology itself. LOGO technology, when not supported by a computer culture built around constructionism and the recognition of the power of learning by designing, has been used in ways that do not reflect the natural learning extended by Papert through microworld design. By creating an *operational folklore* around a microworld, researchers try to create a cultural interface between the natural culture and the technology and, thus, protect the context of learning from being subverted by authoritarian pedagogy.

Examples of operational folklore are numerous. The idea of debugging in computer science converts the idea of "error" into an opportunity for growth. Debugging supports the process of thought becoming systematic, more *stable*. "Playing turtle" or "playing robot" provides a practice to support the use of LOGO or LEGO-LOGO. Granott's research on robot descriptions contributes an operational folklore around the use of electronic bricks in microworlds on self-regulating systems.

10 Conclusion: Constructionist Interactivity

The microworlds discussed in this paper reflect a certain aesthetic of learning environment design, what I call *constructionist interactivity*. Like any aesthetic, it is not a prescription but an evolving set of ideas to guide the design and evaluation of artefacts, in this case learning environments. Four ideas at the core of this aesthetic are objects-to-think-with (e.g. the LOGO turtle, a LEGO® robot), powerful ideas (e.g. recursion, Newtonian motion, cybernetics), tools for designing (the LOGO language, LEGO® bricks) and an operational folklore to support tool use (e.g. debugging, synthetic ethology).

Note 1. Dynaturtle: What follows is a brief description of the LOGO procedures for making a Dynaturtle. The procedure SETUP assigns 0 to the X and Y components of the turtle's initial momentum and a force value of 1.

```
TO SETUP
CG CT PD
MAKE "VX 0
MAKE "VY 0
MAKE "FORCE 1
END
```

NAVIGATOR uses the arrow keys to change the turtle's direction (LT or RT) and to give a kick in the direction the turtle is facing (not the direction in which it is heading). The space bar is used to print on the screen the X and Y components of its current velocity. Notice that the turtle will leave a trace of its movement which is then erased (using CLEAN) when a kick is introduced.

```
TO NAVIGATOR
IFELSE KEY? [MAKE "CHAR READCHAR] [MAKE "CHAR 0]
IF (ASCII :CHAR) = "29 [RT 15 STOP]
IF (ASCII :CHAR) = "28 [LT 15 STOP]
IF (ASCII :CHAR) = "30 [CLEAN KICK STOP]
IF (ASCII :CHAR) = "32 [PRINT SE [The X and Y of resultant: list :vx :vy]
END
```

KICK updates the effect of a kick on the X and Y vectors.

```
TO KICK
MAKE "VX :VX + (:FORCE * SIN HEADING)
MAKE "VY :VY + (:FORCE * COS HEADING)
END
```

DYNATURTLE is the recursive superprocedure that incorporates the use of the other procedures. Turtle moves are produced by adding the X and Y components of the turtle's momentum to its current position.

```
TO DYNATURTLE
NAVIGATOR
SETPOS LIST (XCOR + :VX) (YCOR + :VY)
DYNATURTLE
END
```

Note 2. Animistic Thinking: Granott traces animistic thinking beginning with PIAGET's study of egocentric behavior among children. Young children have difficulty distinguishing between projections of themselves into the physical world and the physical world itself. For example, children have difficulty distinguishing between things thare are alive and not-alive because they view *movement* as a differentiating criterion; a ball is alive because it rolls while a plant is not because it does not. As children identify different criteria for making alive/not-alive distinctions, their answers become more reflective of adult understanding.

In Turkle's study of children's thinking about electronic "toys" (e.g. speak-and-spell, calculators, etc.), younger children do not attribute aliveness to these objects because the toys do not "move". However, older children do attribute aliveness to them because they use psychological rather than physical criteria to make alive/not-alive distinctions. In this virtual world, there are actors taking and receiving actions much like there are in the real world. Because the older children can make analogies between virtual and real worlds, they are compelled to "animate" these transactions (Granott, 1990; Granott, 1991).

References

Ackermann, Edith (1991a). "The Clinical Method as a Tool for Rethinking Learning and Teaching". Paper presented at the annual meeting of the American Educational Research Association. Chicago. April.

Ackermann, Edith (1991b).

Braitenberg, Valentino (1984). *Vehicles. Experiments in Synthetic Psychology.* MIT Press. Cambridge, Massachusetts.

Gargarian, Gregory (1993). *Expressive Intelligence: A Computational Theory of Design.* Media Laboratory Phd thesis. MIT.

Granott, Nira (1990). "Puzzled Minds and Weird Creatures: Phases in the Spontaneous Process of Knowledge Construction." In *Constructionism.* Ablex Publishing Corporation. New Jersey. 295-310.

Granott, Nira (1991). "Machines that love, toys that want: A new look at animism and its possible cognitive function." Unpublished paper.

McGee, Kevin (1992). *Play and Development: The Genesis of Middle-Managers.* Phd thesis. MIT Media Laboratory.

Nettl, Bruno (1983). "Cultural Grey-Out". In *The Study of Ethnomusicology.* University of Illinois Press.

Papert, Seymour (1980). *Mindstorms: Children, Computers and Powerful Ideas.* Basic Books. New York.

Piaget, Jean (1977). *The Essential Piaget: An Interpretive Reference and Guide.* Edited by Howard E. Gruber and J. Jacques Voneche. Basic Books. New York.

Resnick, Mitchel and Fred Martin (1990). "Children and Artificial Life". E&L Memo 10. November.

Wolkomir, Richard (1991). "Working the bugs out of a new breed of 'insect' robots". *Smithsonian.* 22 (3),June, 65-73.

Situation Graphs as Tools for Ordering of Students' Thinking and Understanding of Actual Existing Servo Mechanisms

Jorma Enkenberg

Department of Subject Teacher and Study Adviser Education,
University of Joensuu, 80101 Joensuu, Finland

Abstract. The purpose of this paper is to discuss and evaluate an open-ended LEGO-LOGO prototype project designed to facilitate learning about the structure and function of actual existing servo mechanisms. The theory of cognitive apprenticeship was applied in the modeling of the seventh grade students' learning and their teacher's role. The project was divided into five consequent stages: open investigation of actual existing phenomena, the representation of different forms of actual existing servo mechanisms' structures and functions, the construction of LEGO® models, programming the models with LOGO and writing and presenting reports. The situation graph was used as a tool for knowledge representation and reflection in several phases of the project. In this paper both the methodology and the results of the work done by students will be discussed and evaluated.

Keywords. Cognitive apprenticeship, Control technology, Constructivism, Elementary education, Evaluation, Knowledge representation, LOGO, Longitudinal study, Problem solving, Pupil learning, Pedagogical robotics.

1 Introduction

As technology advances the world in which we live not only grows more complex but at the same time becomes less transparent. The result of using increasingly complex modern technological applications is that the actual work that takes place is to a continually greater degree hidden from those performing or those observing it. As a consequence of this, understanding the construction and functions of servo mechanisms, robots and auto control systems is no longer possible in the same way it was earlier in traditional technology. In order to grasp the principles and functions underlying these knowledge structures, high order skills and proficient performance skills seem to be necessary.

Technological development is challenging our schools to reevaluate both their curricula and goals. The central aims of our educational system should be to develop those values, skills and inclinations in the students that will assist them in dealing with phenomena, both complex and difficult to classify, be they at home, at work or elsewhere. As a result of this the aims, goals, working methods and teacher's role in the learning process need critical reassessment (See e.g. Harel 1991, Resnick 1987, Resnick and Klopfer 1989, De Corte 1990, Collins, Brown and Newman 1989).

Control technology is a significant example of this new expression of technology. For the present it is relatively unknown to most teachers as a learning environment as well as to many computer science instruction researchers. It is rapidly becoming a textbook example of a complex, versatile learning environment where thinking and problem solving processes are naturally emphasized in the learning situation.

Both the theoretical background of the so-called openended Legologo project and its practical applications will be examined in this paper. The goal of the project was to model actual existing servo mechanisms. This took place in a complex learning environment where a projectoriented method was employed; the inner workings of some servo mechanisms were analyzed as a basis for simplified models within the Legologo framework.

One specific goal of the project was the development of a simple, yet broadly applicable tool for thinking. This would support the students' cognitive processes as they planned and carried out the construction of working models which simulated actual existing servo mechanisms and the LOGO programs necessary to run them.

2 From a Simple to a Complex Learning Environment

Researchers currently seem to agree on the concept of meaningful learning. It is characterised as being an active, constructive, cumulative and developmental activity. It is seen as having an initial state and a goal, as does problem solving. The goal is attained by applying various problem solving processes and strategies in the learning situation (See e.g. Shuell 1990, Anderson 1987, Bereiter 1989).

The cognitive theories supporting the concept described above have up to recent times viewed learning as a mental activity, isolated from the ambient culture and taking place only within the confines of the learner's mind. Its connection with the student's life experience, immediate surroundings, readily available technology or parental influence are not held to be significant. Recent research, however, has clearly shown the necessity of anchoring the learning to realistic and authentic situations which parallel the student's experiences and activities. According to the so-called situational cognitive paradigm some demands for change are called for. These include, among others, the time span for teaching goals (45 minutes), the basic unit of learning time (45 minutes), the role of the teacher (information dispenser), the focal point of activity and control (teacher), the choice of the material taught (school knowledge), the essence of knowledge (compartmentalised knowledge), the quality of learning (successful results in a standardised test) and the nature of the learning task (simple) (see e.g. Achtenhagen 1990, Carraher, Carraher and Schlieman 1985 and Rogoff and Lave 1985, Brown, Collins and Duguid 1989).

The following factors have in turn contributed to the changes in the conception of learning: the crises of transfer research (Resnick 1987), the wider acceptance of constructionistic conception of learning (Papert 1990) and the applications of

Vygotsky's social mediation theory as embodied in the so-called cognitive apprenticeship and situational cognition (See e.g. Collins, Brown and Newman 1989, Brown, Collins and Duguid 1989). An analysis of the differences between formal and non-formal learning has also accelerated the demands for changes. Resnick (1987), for instance, has described them in the following way:

- school learning emphasizes the individual character of learning whereas the activities outside school are group oriented.

-school learning is dominated by pure thinking activity without efficient tools. In real life cognitive situations various tools of thinking, such as books, computers, calculators, etc. have an important role.

- school learning uses symbols whereas outside of school the activities are closely connected with concrete objects and events.

- school endeavours to teach skills that are general in nature and for future applications; in real life skills are specific to a given situation.

There are problems in school learning in other areas as well. School goals have traditionally been short term when in fact a perspective that carries the students through for months, years, even their whole lives is needed (e.g. life-long learning). This has moulded the content to be taught in lists, fragmented into small pieces which cannot meet the students' personal needs and skills. One of its manifestations is evaluation which measures only very narrow areas of learning and is largely based on recalling taught material (See Achtenhagen 1990).

Achtenhagen argues we can assume that school learning and its environment should be changed in at least three aspects. First of all, domain specific knowledge should be stressed more than it currently is in the planning of, carrying out of and evaluation of teaching. When planning and choosing subject areas the relationship between the school and the outside world should be considered and emphasized. The relationships between subjects should be apparent on a more profound level. Changes in the real world, external circumstances should be reflected in the reorganization of the learning situation. Newer and ever more complex situations, closer to real life, should be set as goals in the formation of content areas. In order to learn these, new learning strategies must be developed, strategies that would more effectively than ever reflect the demands of the learning environment. The foregoing arguments and demands for change in our view can be focused to deal with a significant part of present school teaching and learning.

Among other things, with the help of the LEGO® TC LOGO system actual existing complex phenomena can be effectively simulated. They can be approached holistically, with all of their essential elements and constituent parts being taken into consideration. The following diagram illustrates those features of the LEGO® TC LOGO system which are significant in the simulation of a real, existing servo mechanism.

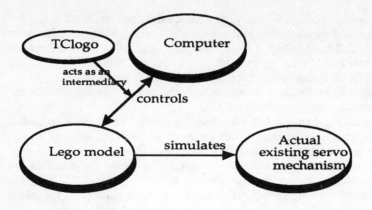

Fig. 1. Modeling of actual existing servo mechanism in LEGOTClogo environment.

In our view LEGO® TC LOGO is a textbook example of a new kind of a complex learning environment where thinking, learning and action are connected with one another in a creative way. The goal of learning is an expression of a modern, technological phenomenon, a real existing servo mechanism. The knowledge embodied in the servo mechanism is the knowledge of an expert, which is not usually apparent nor easily conveyed to the student. In principle the LEGO® TC LOGO system makes this complex knowledge transparent (and easily grasped by) to those working with it.

3 The Developing of Expertise as a Goal of Education

3.1 On Skilful Problem Solving

Research in problem solving over the last few decades has shown that there are still great difficulties in the defining, existence and teaching of domain specific and general problem solving skills. Despite this, our view on the important factors in successful problem solving has been narrowed significantly as a result of in depth research. The following synthesis by De Corte on the prerequisites of skilful problem solving is indicative of the viewpoints that are current among researchers dealing with problem solving.

According to De Corte (1992), in order to be successful in problem solving, among other things the student must be able

- to apply domain specific knowledge including concepts, rules, principles, formulae and algorithms to the situation.

- to use heuristic methods, i.e. systematic search strategies for knowledge in order to analyze and break up problems into its component parts. These heuristics include dividing the problem into more manageable bits (subproblems), depicting the problem in drawings or various graphs.

- to employ metacognitive thinking, i.e. the utilisation of one's own knowledge of the metacognitive processes and carrying out of activities which lead to monitoring one's own though processes. Metacognitive thinking can be connected with the mapping out the problem solving method, reflection on one's own thinking and learning.

- to utilize different learning strategies in order to develop those skills already mentioned.

The above mentioned stresses the difficulties involved in learning those high level skills related to problem solving, i. e. how to effectively communicate thought processes and problem solving skills to the students as part of their education.

3.2 On the Nature of Expertise

Researching skilful accomplishments in the so-called expert novice paradigm has revealed a number of significant differences between the behaviour of the expert and the novice. The differences seem to be related to the skills inherent in problem representation, the nature of the concept evolving from the problem, heuristic strategies arising out of the problem solving situation and the decision-making carried out in demanding tasks. The existence of qualitative differences has been demonstrated in several domains and various areas of expertise (See e.g. Larkin, Mc Dermott, Simon and Simon 1980, Schwartz and Griffin 1986, Glaser 1987).

From the standpoint of skilful accomplishment the most significant qualitative difference relates to problem representation. Where the novice usually focuses his attention on the information directly available from the problem itself, the expert constructs his hypothesis by relating it to domain specific principles and abstract processes. Expert knowledge structure has proven itself to be very goal-oriented and procedural. An expert is able to manage knowledge which is construed as a result of representation. Due to this he can act flexibly when making decisions. Paucity of experience on the part of the novice seems to explain his disability to act in this way. Because of this the novice accomplishments do not become automatic enough and that easily causes a cognitive load to form in the connection of complex information processing.

As can easily be discerned the picture of a successful problem solver is quite similar to the one portrayed by expert research, i.e., someone skilled in accomplishing tasks. In both cases there is a core question of how the learning environment will be organised and how learning and activities will be directed in order to optimally develop high level skills.

This paper hypothesises that it is possible to promote the development of well organised thinking and ordered knowledge by focusing the student's attention on domain specific knowledge and especially on its declarative and procedural components. Various graphic representations can help the student in the reflection and further development of the knowledge structure.

4 Knowledge Acquisition in a Complex Learning Environment

4.1 On Modeling an Expert's Knowledge

Learning that takes place in a complex learning environment can be said to contain at least two significant open questions. First, little information is available on how the learning process should be modelled in order for the student to construe a well ordered view of the target phenomenon. The problem of how to convey efficient learning strategies to the student in order to facilitate the learning of well organised conception of learning is still unanswered.

The second open question deals with the problem of how to make the knowledge of the complex learning target transparent to students. This problem is reminiscent of the one having to do with expert skills, i.e., how to make those high level opaque skills in 'normal' work transparent and facilitate learning them in the school learning environment (See Collins, Brown and Newman 1989, Resnick and Klopfer 1989).

According to the analysis by Shaw and Woodward (1989) in the process of the interpretation of expert knowledge, modeling and the cognitive processes supporting it interact. Figure 2 illustrates this interaction.

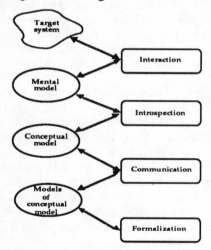

Fig. 2. The modeling processes involved in knowledge acquisition(Shaw and Woodward 1990).

The aforementioned interaction process in our view offers a functional model for the depiction of knowledge development in a complex learning environment. The following short explanation of working in the context of control technology will further illustrate the point.

The beginning phase of working in a complex learning environment seems to emphasize those processes and activities through which the student tries to articulate to himself the perceptions resulting from the interaction and the relationship between them and his own preconceptions. The student classifies situations in line with the mental models he has conceived earlier. If this does not work, he uses the mental model he construed during the interaction. The construing of this model is based on the student's preconceptions or models of the phenomenon or the learning target.

In order for the student to understand a situation which is new and complex, a widening of thought circles is necessary. This can be accomplished through a conceptualisation process, by developing one's own mental model. Modeling takes place with the help of concepts and the result is an improved articulate conceptual model. It can be depicted and communicated in various knowledge representations. The result is a perception which is more intelligible than before and facilitates communication of the phenomenon, e.g. its formalization to oneself and to others.

A given for the formalization of the knowledge relating to the phenomenon is that the model which up to this point has been loosely defined is transformed. Modeling by symbols and language enables the planning and implementation of a computer program which guides the functions of actual existing servo mechanism models.

4.2 The Concept of Cognitive Apprenticeship

In Resnick's and Klopfer's (1989) view knowledge and thinking learned at school can be applied more effectively than before if the subjects and thinking are made more transparent. Consequently, real problems should be chosen for the students to work with at school. Thus learning can take place in a more efficient context and convey to the students the notion that what they learn in school can help them to perceive, interpret and order their own conceptions and experiences.

Transparency, according to Resnick and Klopfer, can be promoted by creating learning situations through which the students can become more aware than before of their own and their fellow students' learning processes, as well as the role of the teacher and his thinking patterns.

The basic principle underlying the theory of cognitive apprenticeship learning (Collins, Brown and Newman 1989) has been to combine the traditional, restricted approach, proven successful in situation based skills, with the modern cognitive learning theory. The method was planned for the learning tacit, opaque and skilful accomplishment combined with expert knowledge and thinking.

In order for opaque knowledge and thinking to be conveyed to students, these must be illustrated and examined in their proper contexts. Learning must, in the main, be based on experience and must emphasize cognitive and metacognitive processes. The goal must be to link learning, thinking and action to actual life situations which can effectively convey the significance of knowledge.

The following teaching strategies are predominant in the applications of cognitive apprenticeship teaching as presented by Collins, Brown and Newman (1989):

Modeling	- is connected to situations where the student follows tasks performed by an expert. Observation enables construing of mental models in certain types of tasks.
Coaching	- has to do with hints and feedback given by the teacher to the student, the goal being to improve the student's performance.
Scaffolding and Fading	- consists of the actions linked to the guiding of the student during his performance. The goal is to get the performance well under way and transfer responsibility for its completion to the student as soon as possible.
Articulation	- refers to the techniques which help students externalize and demonstrate their knowledge and thinking explicitly in connection with their task performance.
Reflection	- leads the students to compare their own mental models and strategies with the teacher's or other students' mental models and strategies.
Exploration	- the goal is to increase the student's independent work and thinking the problems through. It is closely connected with the phases of the learning process, the aim of which is to isolate and define new problems and challenges to be solved.

As expressions of cognitive apprenticeship teaching, Collins, Brown and Newman (1989) mention the so-called reciprocal teaching of learning to read by Paliscar and Brown (1984), Scardamalia's and Bereiter's (1985) method for the teaching of learning to write and Schoenfeld's (1985) method of mathematical problem solving.

In our view the study of the function and structure of actual existing servo mechanisms in a Legologo environment can reflect the principles of cognitive apprenticeship teaching as clarified by the considerations of Resnick (1989) and Collins, Brown and Newman (1989) at least in the following aspects:

- working with concrete and familiar material (Legos) offers efficient possibilities for investigation, experimentation and developing mental models of servo mechanisms.
- working with familiar and powerful learning material which emphasizes explorative activity (Legos and LOGO) can support reflectional thought and thus develop metacognitive thinking skills.
- working in control technology can easily lead to self-regulating and open learning processes which are associated with actual problems and learning targets. This serves to promote the formation of meaning in learning activities.
- the progress of the control technology project can be aimed in the direction of enabling the student to follow his peers' thinking and activities. This takes place when the students are encouraged to observe how other students work and recommending that they discuss issues in groups.

- working with Legos and LOGO can lead to complex cooperative projects which are new for both the teacher and the students. This in turn serves to promote the conveying of the teacher's skilful performance strategies to the students.

To conclude, control technology allows activities to be brought into the classroom that reflect the central principles of cognitive apprenticeship teaching.

5 Situation Graphs as Tools for Ordering Knowledge

The ordering, recalling and reusing of knowledge requires modeling of situations and phenomena. Generally the result of projecting actual problems are their physical or mental models.

The model is considered to be an ideal representation of an actual life phenomenon. Its basis can be analogous or syntactic-semantic. In the first case the model has the structure and function of another model whereas in the latter the mechanism is described by defining syntax and semantics closely related to it (Boy 1991). Here in the investigation of the LEGO® TC LOGO environment both analogous and syntactic-semantic modeling were applied.

Modeling an actual existing phenomenon produces a representation in the form of a text, picture, video picture or graph. The representation can also be a LEGO® model or a LOGO computer program. Its nature is either procedural or declarative. A LOGO program is an example of a procedural representation of the model function.

If the phenomenon in question is described in terms of the concepts characterizing it or being interdependent on each other, the representation is said to be declarative. A semantic net, concept map or a situation graph are examples of expressions of declarative representation. In all the preceding, knowledge is depicted by nodes referring to the concepts and links conveying dependencies between them.

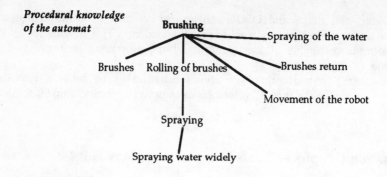

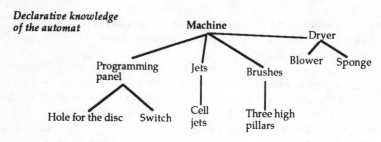

Fig. 4. Procedural and declarative knowledge of automatic washing machine in service station described by situation graph. The graphs was made by a 7 th grade student after spontaneous actual life investigation.

The term situation is used here to characterize the state of a given or perceived actual existing target phenomenon. Three different types can be distinguished: the actual situation, the perceived situation and the desired situation. The actual situation describes the target, the student produces the perceived situation as his interpretation after interacting with it. The desired situation on the other hand describes the goal hoped for.

The simplest version of the situation graph, the tree structure, has been used in this research to order the knowledge inherent in actual existing servo mechanisms. By its aid it is possible to depict the plans for modeling, the operations necessary for its achievement and situations rising out of the modeling process (See Boy 1991). Figure 5 below clarifies the relationships between the various stages of working in control technology environments, the different learning strategies stressed, and representations for knowledge structure.

Modeling   Learning strategies   Tools for knowledge
                                               representation

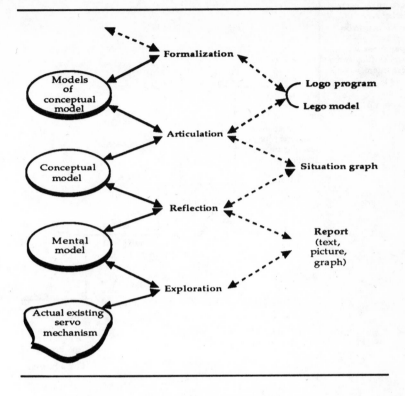

Fig. 5. Modeling, learning strategies and tools for knowledge representation in an open-ended Legologo project.

6 The Open-Ended LEGO-LOGO Project

6.1 How the Project Proceeded

During the project, which took about 25 hours, seventh grade students worked together in pairs. The project itself consisted of five main stages: Open real life investigations, representation of the structure and function of actual existing servo mechanisms (automats) in different representation formats (eg. text, depiction, situation graph), LEGO® model construction, LEGO® model programming and the planning and writing of the report. The situation graph worked in all the stages of the project as a tool for organizing knowledge, for supporting planning and backward oriented reflection. The stages, their central contents and the average time spent on them are shown in figure 6.

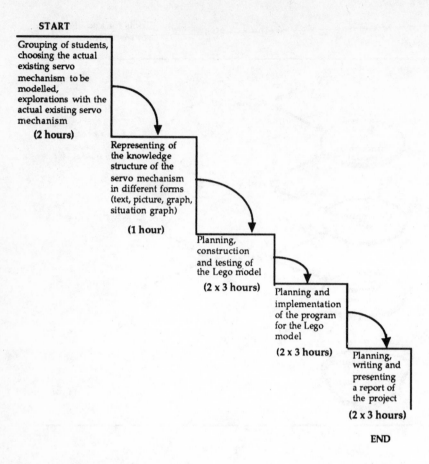

Fig. 6. Stages and their duration in an open-ended Legologo project.

So-called actual existing servo mechanisms were chosen as objects to be modelled. Examples of some of these were a lift, an automatic door, traffic lights, a bottle return automat, parking building gate, a coin operated soft drink automat and an escalator in a department store.

The students were allowed to explain and model which servo mechanism they wanted, the only limitation being that no two pairs could work on the same project.

First the students became acquainted with how the actual existing servo mechanisms performed in practice. This was done by analysing and probing how they worked and what parts they had to have in order to accomplish their specific functions. Next the students wrote reports on what they had construed based on their investigation. This report served to support the students' so-called process of 'backward oriented reflection'. It was also hoped that these reports could be used later in evaluating the students' preconceptions and the mental models created during their investigation. When they were working on their reports, the students were asked to reflect on both the structure and function of their respective servo-mechanisms as well as to describe them meaningfully.

The next stage was for the students to organize their thoughts in tree structure form. This done, the planning of the models was begun. The students were also asked to consider which LEGO® parts they would need to build their LEGO® models. The available building materials consisted of various LEGO® blocks of different sizes and shapes, motors, necessary wires, sensors and lights. Immediately after this the students were asked to depict on paper how they would explain to a friend, who knew nothing about the servo mechanism and or the model in question. Then they were to draw on paper how the model would be built. The underlying purpose of this exercise was to encourage the students to do some preplanning before starting to build their model.

In the next phase, the actual building of the LEGO® model was undertaken. Because the students were used to following printed LEGO® building instructions, some difficulties arose when they had to build their own from scratch. The process turned out to be a great deal more demanding than expected. It should be mentioned that almost every group at least once reached an impasse and had to dismantle their prototype. This was due to unforeseen difficulties in either the preconstrued plan or conception of the model.

When the students felt their model was ready the next step was to write the LEGO® TC LOGO program. While this phase proved to be the most challenging, it was also the most interesting. As the programming was not easy, the teacher was repeatedly required to lend support to the students' thinking processes. Before the programming started the teacher reminded the students of the washing machine tree structure diagram and the listing of program procedure which they had studied earlier (see figure 3.). LEGO® TC LOGO vocabulary and commands were also reviewed. These would be required to run the motors, lights and sensors of the models. The topdown method was used in the planning and carrying out of the program. The situation graph was also used as a tool to guide thinking processes.

Once a program functioned and was ready and the students felt it was good enough, the planning and writing of reports was started. The reports were presented in the form of a slide-show using the designer program.

The project ended with the publication of the results which in turn were submitted for the comments and evaluation of both fellow students and experts.

The following is a summary of the different phases of the project and what was thought to be meaningful for the development of the students' thinking and knowledge.

Spontaneous Investigation   - examines the students' conceptions of servo mechanisms.
- solicits new ideas and views on servo mechanisms.
- aids the understanding the structure and function of the servo mechanism.
- creates strong ties between the context and situation.

Reporting	- develops and cultivates the students' conceptions and thinking. - obliges the students to reflect on their own thinking back to the servo mechanism once more. - facilitates the formal modeling of the structure and function of the servo mechanism which is to take place later.
Situation Graph/ Tree Structure	- models the function and structure of the servo mechanism. - the results are explicit and decontextualized forms of presentation of the model structure and function. - these are the plans for the building of the LEGO® models and the programming of their functioning. - the graphs depict the relationships of the model parts (declarative knowledge) and clarify how the process advances (procedural knowledge). Graphs help the student to order his thoughts.
Building of the LEGO® Model	- the student must reflect on his thinking of the servo mechanisms. To be successful, the building requires planning, reflection (about the actual existing servo mechanism, the preconception of it and of the tree structure), articulation of the problem, use of spatial thinking, experimenting and developing of one's own thinking.
Programming the Model	- the student must reflect and correct his conceptions of the function of the model, plan a program based on these reflections and conceptions, divide the problem into subproblems, experiment and develop his plans, recount his knowledge in a procedural form and solve demanding problems.
Portraying Results	- the student must depict his thinking in various representations (text, picture and graphs and communicate his ideas to others in public). This can be seen as a process which promotes the eliciting and developing of conceptions.

6.2 The Role of the Teacher during the Project

The teacher aimed to apply the strategies of cognitive apprenticeship learning when guiding the students through all phases of the project. The goal in doing this was to further the transparent accessibility of both the teacher's and the students' thinking. The following is a short outline of how the strategies were present in the connection of various activities.

The teacher ventured, more often than is the norm in traditional classroom instruction, to think out loud and explain his own ideas, problems involved and recommendations for solutions. This would take place, for instance, when the teacher would start discussion of the project from one particular aspect or in guiding the students in the programming.

The direct teaching strategies, modeling and coaching of cognitive apprenticeship learning were applied especially at the beginning of the programming and in the planning and writing of the reports. These phases were very demanding for the students and seemed best suited for direct teaching. The realization of the students' own conceptions and plans was encouraged by indirect strategies in the making of the tree structure, building of the LEGO® model as well as in the programming. This was apparent in the openended nature of the project. The students' own spontaneous experiments with their impasses were encouraged in their activities, e.g. in the building phase of the LEGO® models. The teacher also supported the students actively. However, in discussions with the students, it was hoped that they were expected to both reflect and express verbally their thoughts and situation related problems to something they had done or examined earlier. The tree structure was repeatedly used as a tool for the reflection. Questions such as those below were often used to guide the thought processes and goal setting:

- "Please explain what you're doing."
- "What is the main thing (goal) you're trying to get done here?"
- "Please tell me how you think your model should do what it's supposed to behave?"
- "Can you explain to me what you've done up to now?"
- "Can you break your program down into steps for me?"
- "How can you change your program to make it do *this* and *that*?"
- "Please explain how your program works, right from the very start, to the very end."
- "How far have you got to with your program?"
- "How would your program work if you changed where these procedures are to some place else?"
- "You remember the tree structure we looked at before? How is your program similar to it?"
- "How would you explain to a friend who doesn't know anything about your project, how your model works?"

In general the students were encouraged to share their ideas, actively experiment with their plans and compare their problems or solutions with those of other groups. They were also encouraged to move freely around in the classroom.

7 Concluding Remarks

It is not within the scope of this paper to report in detail how the students actually carried out their projects, through their various stages and accomplishments. The following is an assessment, without going into detail, of how the goals set for the LEGO-LOGO project were achieved. The analysis of the applicability of cognitive apprenticeship learning and of the situation graph is of primary importance here.

Throughout the whole project the students worked diligently and responsibly. They were very involved in what they were doing most of the time. In general the students had no trouble working with LEGO® blocks and the LOGO program. Nine student pairs out of ten were able to build their model and program it within a reasonable time span with the help of the teacher. Even those students who had great difficulties and were in need of the teacher's scaffolding and coaching to stay with the project seemed to enjoy working with the Legos and the LOGO program.

The students seemed to be deeply involved and on several occasions were genuinely engrossed in their work. This is apparent from the fact that the teacher had trouble getting the students to take breaks during the threehour working period. One reason for this may be that the level of difficulty of the project may have been optimal to the students' skills and inclinations. Moreover, the strategies the teacher chose to use, consistently supported the progress of the work. The learning environment itself could have had an effect of the advancement of the dynamics of the project. The LEGO® TC LOGO environment provided ample opportunities for the students to develop their own spontaneous ideas and test their effectiveness.

The use of the tree structure in the ordering of the students' thoughts proved to be successful even though some students felt it to be unnecessary and superfluous. At the beginning of the project the tree structure seemed to be useful in the organising of the students' knowledge. This was apparent in the reports they made based on their spontaneous investigations. It also was an efficient tool in the programming of the model with the LOGO program as well as in the planning of the report at the end of the project. During the actual building of the LEGO® model the tree structure was not needed very much while it did seem to be a background factor in guiding the students' interest. In the LEGO® building phase the tree structure was examined when the students came to an impasse. With it a well organised picture of the main structural elements of the model and their relationships was conveyed.

Cognitive apprenticeship learning strategies seemed to perform relatively well in a complex learning environment. The strategies of scaffolding and fading modelled the teacher's activities well when it was time to transfer responsibility to the students. Reciprocal teaching, the reversal of the roles of teacher and students in the reflection phase appeared to be a useful way to emphasize the strategic views in each student pair's and teacher's problem solving situations.

Exploration and articulation were stressed in the students' work. One method often used and closely linked to these was the conscious utilization of impasse situations. The teacher was unable to help all the students all the time even though he would have liked to. The students were asked to pause and reflect a little more on their problem. During the pause they were also encouraged to move about and observe what their fellow students had accomplished. This seemed to be an effective vehicle for the articulation of one's own problems. Delays in receiving assistance and support appeared to develop patience in the students in meeting difficulties and potential impasse situations.

Finally it can be stated that the LEGO-LOGO project, carried out in the context of applying some of the central principles of cognitive apprenticeship learning was able to convey an ordered view of the nature of actual existing servo mechanisms and the so-called transparent knowledge within them. A great influence here naturally was the LEGO® TC LOGO environment itself. However, the strategic learning methods linked with depicting the knowledge of actual existing servo mechanisms graphically and with the aid of tree structures, worked as significant tools for thinking. The use of these can be said to promote the ordering of the conceptual model built based on authentic existing servo mechanism and fostered communicating of their essential ideas.

In our view the methods of cognitive apprenticeship learning are useful in a wider context in examining and learning about complex modern technological phenomena and their applications.

References

Achtenhagen F. (1990) Development of problem solving skills in natural settings. In Carratero M., Pope M., Simons R. and Pozo J.(Eds.) Learning and Instruction. European Research in an International Context: Vol III. Pergamon Press.

Anderson J.R. (1987) Skill acquisition: Compilation of weak-method problem solutions. Psychological Review, 94, 192-210.

Bereiter C. (1990) Aspects of an educational learning theory. Review of Educational Research. 60 (4), 603-624.

Boy G. (1991) Intelligent assistant systems. Knowledge-based systems. Vol. 6. Cornwall: Academic Press.

Brown J.S., Collins A. and Duguid P. (1989) Situated cognition and the culture of learning. Educational Researcher, 18 (1), 32-42.

Carrager T.N., Carrager D.W. and Schlieman A.D. (1985) Mathematics in the streets and the schools. British Journal of Developmental Psychology, 3, 21-29.

Collins A., Brown J.S. and Newman S.E. (1989) Cognitive apprenticeship: Teaching the craft of reading, writing and mathematics. In Resnick L. (Ed.) Knowing, learning and instruction. Essays in honor of Robert Glaser. Hillsdale, NJ: Lawrence Erlbaum Associates Inc.

De Corte E. (1990) Towards powerful learning environments for the acquisition of problem-solving skills. European Journal of Psychology of Education, 1, 5-19.

De Corte E. (1992) On the learning and teaching of problem-solving skills in mathematics and LOGO programming. Applied Psychology: An International Review, 41 (4), 317-331.

Glaser R. (1987) Thoughts on expertise. In Schooler C., Schaic W. (Eds.) Cognitive Functioning and Social Structure over the Life Courses. Norwood: Ablex.

Harel I. (1991) Children designers. Norwood NJ: Ablex

Larkin J., McDermott J., Simon D. and Simon H. (1980) Expert and novice performance in solving physics problems. Science, 208, 1335-1342.

Paliscar A.S., Brown A.L. (1984) Reciprocal teaching of comprehension-fostering and comprehension-monitoring activities. Cognition and Instruction, 1, 117-175.

Papert S. (1991) Situated constructionism. In Harel I., Papert S. (Eds.) Constructionism. Norwood, NJ: Ablex.

Resnick L. (1987) Learning in school and out. Educational Researcher, 16 (9), 13-20.

Resnick L., Klopfer L. (1989) Toward the thinking curriculum: An overview. In Resnick L., Klopfer L. (Eds.) Toward the Thinking Curriculum: Current Cognitive Research. 1989 ASCD Yearbook. ASCD.

Rogoff B., Lave J. (1984) Everyday cognition: Its development in social context. Cambridge, MA: Harvard University Press.

Scardamalia M., Bereiter C. (1985) Fostering the development of self regulation in children's knowledge processing. In Chipman S., Segal J. and Glaser R. (Eds.) Thinking and Learning Skills. Vol 2. Research and Open Questions. Hillsdale, NJ: Lawrence Erlbaum Associates.

Shaw L.G., Woodward J.B. (1990) Modeling expert knowledge. Knowledge Acquisition, 2, 179-206.

Schoenfeld A.H. (1985) Mathematical problem solving. New York: Academic Press.

Schwartz S., Griffin T. (1986) Medical thinking: The psychology of medical judgement and decicion making. New York: Springer.

Shuell T.J. (1990) Teaching and learning as problem solving. Theory into Practice, 29, 102-108.

Vygotsky L.S. (1978) Mind in society. The development of higher psychological processes. Cambridge, MA: Harvard University Press.

Problems Associated with Getting Control Technology Working in Schools

Reg Eyre

Wiltshire Education Authority, Trowbridge, BA14 8JB, United Kingdom

Abstract. This paper outlines the writer's experiences and difficulties with making control technology an acceptable and accessible activity. It aims to show how developments in hardware, software, training of teachers and national curriculum reform have had an effect on the current practice seen in Wiltshire schools.

Keywords. Assessment, Control technology, Elementary education, LOGO, Pedagogic attitude, Problem solving, Pedagogical robotics, Sensors, Technology learning, Teacher education.

1 Personal Experience

I joined the staff at the College of St. Paul and St. Mary in Cheltenham in 1982 with a brief to expand the In-Service Training (Inset) of teachers in the local area with respect to classroom practice in the use of microcomputers.

The most frequently used computers in schools at this time were Commodore Pets, Research Machines 380Zs, the BBC model B and a variety of enthusiast built kit machines.

Since the Local Education Authorities had decided to standardise on the BBC computer, I decided that that both initial teacher training and Inset should concentrate on these machines.

My colleague Allan Philpot, of the Science Education department, had seen the potential of using the BBC computer's various interface connections for developing work with sensors to make science experiments more exciting and immediate. Our initial collaboration concentrated on the writing of small pieces of code to capture information read from the Input/Output (I/O) and Analogue/Digital (Adval) ports and present this in tabular or graphical formats.

We changed our courses as more suitable and reliable equipment and software became available. Over the same period of time, national government initiatives had made funding available for a variety of Inset course formats to be tried and evaluated.

The Education Reform Act of 1988 has led to a nationally specified curriculum which includes both the use of Information Technology (IT) as a curriculum delivery mechanism for all subjects as well as some aspects of IT as worthy of study in their own right. Since schools are now obliged, by law, to provide access for their pupils to Control Technology as part of their National Curriculum entitlement, there is now a change in emphasis in Inset provision and teacher attitudes to this aspect of IT.

I left the College in 1990 to work for Wiltshire Education Authority as an Advisory Teacher, so that I would have more opportunities to develop Inset work and become closer to the work in classrooms where effective changes in classroom practice can be made. In the past two years, the IT Team for Wiltshire has put together a comprehensive package for teachers covering hardware, software, documentation and Inset. The documentation covers aspects of implementing IT through each of the national curriculum subjects, recording and monitoring pupil progress at all stages in their IT entitlement and progressions of activities in IT across the the curriculum for the primary age range.

2 Hardware

As explained earlier from 1982 we standardised on the BBC computer. This machine has the benefit of having several interface ports available as standard and hence peripheral manufacturers supply equipment which can be used immediately without having to worry about buying and fitting other internal modifications for the basic computer.

The BBC computer is now out of production but replacement parts are plentiful so that its continuation as a general purpose machine is assumed for the next few years. We continue to use it and promote it in our schools because it has an extensive software base and can be used in all areas of the curriculum. We have looked at other computers but remain convinced that the BBC computer is still the best value-for-money machine for control work but also for all other aspects of the curriculum, since schools have limited funding and many cannot afford to have machines dedicated to one use only.

In 1982 we were using small interface boards designed by enthusiasts and made by Allan Philpot at the College. These were connected into the I/O port and set up for four inputs and four outputs via one millimetre jack sockets. These were very fiddly for adults and children to use, and made all control work appear very Heath Robinson-ish. Similar home made boards were designed, made and used to connect to the Adval port as well as motor reverse boards. Using this primitive equipment we ran Inset courses and worked alongside teachers in the local area with their classes of both primary and secondary age groups.

The first commercially available interface we used was the Cambridge Control Buffer Box System. This was expensive, well made and comprehensive in what it offered. The package included an input box with room for eight inputs connected to the I/O port, a mains powered and protected output box connected to the printer port with room for eight outputs and provision for pairs of output ports to be connected for motor reversing and a four input Analogue/Digital box connected to the A/D port. The package contained a range of sensors including a tilt switch, pressure mat, light sensor,etc and some output devices such as motors and lights. Input devices were distinguished from output devices by the use of 2.5mm jack plugs for inputs and 3.5mm jack plugs for output devices.

We eventually recommended the use of the Deltronics package which was similar but much cheaper. Motor reversing circuitry was built into the buffer box and the connections to the computer were the same as the Cambridge boxes. The other small advantages were that the Deltronics kit was contained in one box instead of two and that software developments had led to screen graphics that showed a display of the connections corresponding to the Deltronics layout.

We have not considered the LEGO® control system because our experience has been that children want access to as many input and output connections as possible and the LEGO® system only allows two inputs and six outputs.

We have always encouraged teachers to make their own connections and digital sensors. The best set of connections are the two types of jack plugs with one metre of wire and screw terminals at the other end. These allow both teachers and children to design and build their own versions of sensors such as tilt switches and move them away from the fixed ideas and limitations of the commercially available products. Another advantage of this approach is that children can become less restricted in thinking about what they want to control and how they might sense position in the control of what they have envisaged.

More recent developments are concerned with the use of analogue sensors. We have tended to give these a low priority so that teachers could focus on the main aspects of control and the use of digital sensors. Our view is that the more experienced teachers are able to see the potential of using a variable input against a simple on/off switch. The developments concerning A/D interfaces and software have concentrated on making this a stand-alone device for use in science experiments where we are trying to incorporate such use into control work.

3 Software

In 1982 we were having to use routines written in BASIC since this was the resident language of the BBC computer. We were concerned about the use of such code and how it should be explained, introduced and used by children. For example:

```
10 ?65122 = 15
20 ?65120 = 0
30 FOR TIME = 1 TO 500
40 PRINT "THE SWITCH IS OFF"
50 NEXT TIME
60 ?65120 = 1
70 FOR TIME = 1 TO 2000
80 PRINT "THE SWITCH IS ON"
90 NEXT TIME
100 GOTO 20
```

The children were told that this code would switch their buggies on and off and were asked to investigate how far their buggies travelled, and if they could modify the code to make their buggies travel one metre or for 5 seconds and stop.

We considered a change to LOGO when it became available on a ROM chip for the BBC computer. We were even more fortunate when various freely available extensions to LOGO became available; we could then ignore switching outputs on and off by direct addressing and using newly defined primitives such as SWITCHON and SWITCHOFF which were much more meaningful. We decided to use the Control LOGO package that became available from the Primary Section of the Micro-electronics Education Programme (MEP). This had a friendly graphical presentation on the screen and allowed full use of the LOGO programming language. Some teachers of older secondary pupils who came on our Inset courses were happy to start from the base of using these logo extensions and working back to programming for control using only raw LOGO. They claimed that this would give their pupils insights into basic machine architecture.

The purchase of LOGO chips for BBC computers is still a large expense for many small schools in Wiltshire and an alternative has become available from the National Council for Educational Technology (NCET). (This is the body which took over from MEP when their programme came to an end.)

NCET have produced and published a control package called Contact. Contact is written in a combination of Basic and machine code and is made to look and perform like Control LOGO which was available from MEP. The main advantages are that it is cheap and works in a similar way to Dart which is a stripped down version of the turtle graphics aspect of LOGO. The disadvantages are that there are limitations in what can be done, such as simple arithmetic, and that there is no access to other features which using the full version of LOGO possesses.

Various other control languages, such as Legolines, have been considered and discounted because they do not add much to the process of understanding of what is happening. They appear to be Basic-like or LOGO-like in ways which are not as friendly as Contact or Control LOGO.

The Contact package contains not only the software but also materials developed from the MEP and NCET training courses for teacher trainers. Ideas are given for classroom management, work with construction materials, simple ideas for using electricity and suggestions for implementing a policy for control technology in the primary school.

4 Construction Materials

When we started work with teachers and children we used cheap supplies of offcuts of wood, wire from coathangers and a variety of cleaned household 'junk'. Since then we have used many construction kits and even more 'junk' from a variety of sources. Since both construction kits and other materials have their advantages and disadvantages, we let teachers decide what they wish to use when they are working with their pupils in their own classrooms. The Inset courses allow time for teachers to work with as great a variety as possible to allow them time to make decisions on suitability for their own environments.

The construction kit we use most frequently is LEGO® Technics, since LEGO® is commonly found in most schools and homes.

The advisory team for technology in Wiltshire has produced 'A Guide to Using Construction Kits in Schools' which documents all the commonly available kits with a description of each kit, the technical and scientific areas that can be explored using each kit and some photographs of Wiltshire school children using each kit for various topics. The booklet also contains some teaching ideas and issues to be considered when using construction kits in the classroom.

Typical of the questions we raise for course members on our Inset courses to consider are the following:

- Are construction kits available to all the children in your school?
- Are these kits aimed at the right level for the children at every stage?
- Do you have a progression of activities and use of construction materials?
- If you intend to use associated work cards, have you considered the language and vocabulary?
- Are all staff in your school aware of the above issues? I.e. Is there a school policy covering the use of construction materials?

An advantage when using LEGO® in the classroom is that many children have access to it in the home, so teachers can generally pass the 'play' stage (intended for becoming familiar with the way it can be used) and can therefore progress immediately to meaningful activities and investigations which make use of it.

The use of 'junk' materials is full of problems. Children will be confronted by a random variety of shapes which do not naturally fit together, there will appear to be no easy way of connecting one piece to another and each piece has different properties to any other. The teacher's role is to guide the children, via the development of manipulative, scientific and problem-solving skills, to an understanding of how the materials are to be used.

5 Types of Inset

The first courses we ran were pioneering efforts with only passing reference to classroom practice and expectations. The use of computers in schools was still reasonably exploratory and the software base was growing rapidly with a mixture of good and indifferent programs.

Teachers attended after-school courses and were introduced to 'design and make' activities which led naturally to the control of their machines from the computer. We fed back our own experiences of working with children in local classrooms into these courses so that ideas were progressed as to what might be considered to be good practice.

In general, control technology was fed into courses on science, practical problem solving and LOGO rather than making control technology the focus of the whole course.

At this time, teachers could get funding and release for twenty day courses. I developed a 20-day LOGO course with our Local Education Authorities. The aim was to produce a coherent progression of LOGO activities and experiences for children from primary to secondary school. One aspect of this course was an intensive three day session on control technology.

Wiltshire Education Authority maintains a policy of not just supplying equipment to schools, but insisting that teachers receive the necessary equipment together with a training course. For control technology courses this was to be a five day intensive residential course.

This five-day course had to cover:

- Work on electricity including circuits, conductivity and switches.
- Learning to solder.
- Using LEGO® to find out about gears and gearing.
- Using the Deltronics buffer box to control outputs.
- Learning how to program for outputs and inputs.
- Making switches to be connected to the buffer box to control outputs.
- Learning techniques for using wood.
- Learning about the design process cycle.
- Discussing issues of safety, classroom management and implementation.

These topics were integrated around a series of projects designed to give practical experiences in each of the newly learned areas.

The teachers received the equipment for their schools which they then used on the course. This consisted of:

- Sets of LEGO® Technic 1 and 2.
- Deltronics Buffer box.
- Class set of tools.
- Contact software pack.
- Other useful materials such as wood, wire, magnets, etc.

After the course, teachers are expected to know what they have and how it can be used in their own classrooms.

Further support is available from the IT Advisory Team in the form of a telephone call or, in more interesting cases, in the form of a visit to the classroom. We have also tried the idea of 'Control Challenges' where the teachers bring small groups of children from their classrooms to a venue to show their accomplishments to other teachers and their pupils.

Money for training courses has been reduced from central government funding and we have therefore had to drop the residential aspect of our control courses. We still feel that the residential aspect was important and effective since teachers could give all their attention to the course and worked well in excess of our expectations, for example, getting up very early or staying up very late to finish a project.

From this year, the control courses are to be five half-day/evening sessions. This means that the teachers will have worked the morning session at school and travelled to the course venue for a 13.00 hours start, finishing at 19.00 hours. The course programme will be a compressed version of that outlined above for the five-day residential course, and with fewer sub projects.

The most effective and costly Inset is when one of the IT Advisory Team goes into a classroom and works alongside a teacher with their class to show how such work can be developed in that teachers own teaching environment with only the equipment that is found in that classroom. Knowing that this style of working is impractical, we are trying different forms of Inset but the major problem remains one of knowing that effective change of practice has been achieved.

6 Attitudes and Confidence

There is no doubt that teachers who have been on a residential control technology course feel that they have had a valuable experience. Our aim was to give them a rich environment in which they to became totally involved with all the processes, so that the newly learned techniques would translate into a change in classroom practice.

Most of the course leaders running this type of course have found that teachers go back to their schools enthused enough to try out their new skills, but that this euphoria generally lasts for only about a year. Economically speaking, these courses are expensive to run and, if the resulting change in classroom practice is as slight as we believe, the justification for continuing them is called into question.

It may be that all forms of Inset only have a brief effect in terms of change in practice, or it may be that it is the nature of these particular courses with their associated content needs to be analysed further.

One of the issues which constantly arises on such courses is the lack of knowledge and experiences of mechanisms. In practice, this means that when teachers are asked to make a railway barrier for example, we find that they have never really observed the mechanisms involved and also appear to lack earlier experiences which might help them to guess at what the mechanisms might be. Questioning about their initial drawings (as to how such a mechanism might work) reveals a lack of knowledge of the effects of gravity, friction, bearings, pivots, counter-weights, control or sense devises and fail-safe mechanisms. Because of this lack of background experience, we find that teachers are at an early stage of experimenting with materials and their properties. It may be that their own unease concerning this is such that they do not want to allow themselves to be 'caught out' by the pupils in their classrooms.

Our courses can only provide a limited amount of time for such teachers to come to terms with their own shortcomings but it can seriously undermine the approach we would like them to take. We want to encourage teachers to use the "What ... if ....?" approach with their pupils which should help pupils to experiment for themselves. To adopt this approach, however, does require a degree of self-confidence so that the pupil and teacher working together might find better solutions to problems. It also means accepting that, for many situations, new solutions may be worse than existing ones. This is part of the design process which is written into our National Curriculum. Many teachers express unease at questioning their pupils in this way because they feel they ought to know the 'best' solutions, somewhat like always knowing the correct answer to a mathematics problem. They find it hard to accept that there is often no 'correct' solution, only poor, reasonable, good and better solutions to many problems.

Having built a mechanism to do a particular job, it is interesting to note that both teachers and pupils form a reluctance to change the actual mechanism when problems arise in the control phase of development. Programming solutions are often sought to problems which arise, instead of either reviewing the actual mechanism with a view to changing its characteristics, or even scrapping the mechanism and starting again. Programming solutions often rely on timing which ignores the effects of gravity or friction inherent in a mechanism. For example, in making a model lift which is required to stop at various floors, we find that the motor controlling the lift is operated by a time period and shock is registered at the fact that the time period for the lift to go up one floor is longer than the time to come down one floor. This is a good point at which to explain the effect of gravity! Having put in appropriate time periods, the second realisation is that the lift does not consistently stop at the required points. We use this situation to introduce the value of sensors to indicate position.

The railway barrier problem is similar to the lift problem except that one can use gravity to lower the barrier in the event of power failure which introduces the notion of fail-safe mechanisms.

Our aim is always to try and develop a feeling for simplistic mechanisms, i.e. to do away with unnecessary complications.

We are beginning to feel that modern society is working against this principle. Modern cars now abound with systems which cannot be tampered with by the home mechanic. If something is not quite right with the car, it has got to go to a franchised mechanic who has diagnostic equipment to check where the fault might lie. In the past, a competent owner could make adjustments to various components and test the results to see if an improvement had been made. The older the car, the easier it is to make such adjustments to improve the running of the car. The contention is that with more of the artifacts around us being made of non-adjustable parts the fewer the opportunities that exist for us to experiment with mechanisms and hence we will lose the natural curiosity to want to know how mechanisms function.

7 Agents For Change in Classroom Practice

All of the experiences outlined above were in use before the Government implemented the National Curriculum which specifies the content of what has to be taught in English and Welsh schools. Control Technology is one of a continuum of activities in the Control and Measurement strand of the IT Capability Attainment Target which, in turn, is part of the Technology section of the National Curriculum. Being such a minor part of the whole curriculum, there is a concern that such work will be ignored if the relevant staff expertise is not available within a school.

Before the National Curriculum was in place, many of the people involved in the introduction of new technologies through the In-service education process hoped that enthusiasm and involvement in the exciting applications of Control Technology would inspire teachers to do lots of creative work based on the ideas we were putting forward. This does not appear to have happened. Many teachers are responding to the introduction of the National Curriculum and are finding that there is too much specified content. This means that many are loath to experiment with new ideas which are not in the main-stream of curriculum content. The fact that Control Technology is part of the specified content means that those with a responsibility for Inset can use more than gentle persuasion to coerce some change in classroom practice.

Another agent for change might be the children themselves. The natural progression for activities with materials starts with experimentation with soft materials such as sand and water and works through to hard materials such as wood and plastic. The making of objects using such materials often leads to a desire for movement of these models which involves experimentation with different power sources such as rubber bands, wind or electric motors. This is often followed by a desire to control such movement. This is the process which I had previously hoped would cause teachers to implement ideas for using Control Technology.

The other major agent for change is pressure from society. The problem in this case is that parents are being told by the press and government that teachers are not even teaching the basics properly so the teachers' efforts are not being directed at the whole curriculum, including Control Technology, but at those aspects which are being publicized as being of most importance.

Some advantages that have come from the National Curriculum include the fact that more work is being done in science and technology so that we can now assume basic knowledge of simple electric circuits and more experience with 3-D model making. Our courses can now build on such experiences with relatively few additional concepts to make control technology happen in our schools.

Promoting Active Learning:
A Pragmatic Approach

Chris Robinson

Horndean C. E. Middle School, Five Heads Road, Horndean Waterlooville, Hants PO8 9 NW, United Kingdom, Tel. (44) 705 592236

Abstract. Many teachers follow a didactic style of teaching which may be inappropriate to developing the child as an active learner. This paper describes some of the many problems associated with promoting active learning environments and explores some strategies that have been employed with varying degrees of success of failure.

Keywords. Active learning, Computers, Constructivism, Control technology, Curriculum, Elementary education, Learning environment, LOGO, Progression, Resources, Teaching methods.

1 Introduction

Creating "Active" learning environments is something I have experimented with throughout my teaching career.

When I was trained in the sixties, the Nuffield foundation quoted the Chinese proverb:

"If I hear, I forget. If I see, I remember. If I do, I understand."

I remember from my own schooling the excitement of practical science lessons even if I did not understand what the experiments were about when we were asked to "Find Young's Modulus" and although I never got the "right" answer.

During my career as a teacher, I have recognised a paradox: children learn in spite of school; they often learn little because of school. As an illustration of this, is an observation concerning the introduction of concepts of electrical circuits with children carried out during three years as an advisory teacher.

Children starting school have usually been used to close contact with a parent or other adult receiving individual attention and believing they were the most important person in the world. It comes as a shock to find another twenty children believing the same thing.

The reception teacher often finds her most important role is to teach co-operation. Introducing electricity can help. I provide one child with a battery and another with a cheap motor. (They have insufficient power to be harmful.) If they co-operate, two children soon discover how to make the motor work. (It is also a useful way to introduce teachers to using electricity!)

At this point, I ask the children if they can make the motor turn the other way. Most five year old children look at me as if I am stupid and reverse the wires.

Using exactly the same approach with eleven and twelve year old children who have never had the opportunity to experiment with electricity before, not only do they find it harder to co-operate in the first place but when challenged to attempt to reverse the motor, they seem to find it one of the most difficult things to do. Could it be they think it is a trick question that is too easy for them or is it a fear of being "wrong"?

I have the idea that, if we are not careful, what we do in school is teach children not to learn. What we actually do is inhibit them.

Consider the ways we have traditionally taught children the four rules of mathematics. I am quite certain, if I am representative of most adults, that we do not use those methods ourselves when confronted with mathematical computation. Did schooling actually help us acquire the methods we use, and could they have been taught more efficiently?

These are the issues which have convinced me of the importance of methods of education in which children play an active part in their own learning.

2 Computers and active learning

For many years I taught my own class behind my own closed door thinking other teachers worked in the same way as I did until the advent of computers. I found they fitted perfectly with my preferred teaching method and was surprised when I discovered other teachers had difficulty fitting them to their teaching styles. That was when I realised I had a lot of work to do with other teachers.

I started by telling other teachers what they could do with the computer and getting quite excited and got nowhere.

I was asked to talk to teachers from other schools and found one or two who began to share my enthusiasm. Was it a case of "a prophet is not recognised in his own home", or was it they shared my preferred teaching methodology? If that was the case, how could I convince other teachers to adopt the active learning approach?

3 Strategies to promote active learning

Teachers are quite happy using the methods which they have used for years without asking how effective those methods are. Other strategies have to be adopted.

3.1 Show it is possible

As an advisory teacher, I would visit classes where I would work alongside another teacher. Frequently I would provide an exemplar lesson to start a project and return a few weeks later to observe that teacher using a similar approach towards the end of the project having been accessible via telephone, and intermediate visits to other teachers in the same school, in between.

This strategy was reasonably effective because it removed the argument often provided by cynical teachers on courses that, "It's all very well for you to say but you haven't got the problems of my class." (or room or colleagues or headteacher or resources etc).

Unfortunately, however easy I know Control Technology is, it can very soon start to look complicated with wires everywhere and, although the children may have a very good idea of what is going on, their teacher who has not been with them all the time because she has others to look after as well, does not and uses the excuse, "It's far too difficult for me," which complements the excuse, "We have not got the equipment."

3.2 Produce pieces of writing

Another approach I use is to produce pieces of writing. I had magazine articles and booklets published but never really knew how useful they were or whether they were even read. I discovered later that one article I had written involving music had a fundamental error in it. The fact I never received any correspondence on the matter suggests to me few can have read it.

Anyone having acquired a new camera, video recorder, fax machine etc will know how much easier it is to be shown how to use it rather than read the manual which is only used in the last recourse when all else fails.

In considering what writing could be useful, the British Logo User Group produced some "crib sheets" giving sufficient basic knowledge to enable teachers to feel confident about using LOGO. I have seen them given to children to learn Logo from by working through them as if they were text books - which is not what they were intended for.

The problem with providing written help is there is so much of it. All primary teachers in England and Wales receive a minimum of nine ring binders full of national curriculum plus additional support materials, "non-statutory guidance" and the like, filling one metre of shelf space. The Local Education Authority produces more of it. Some product manufacturers, being "helpful" produce more. School policies have to encompass these documents and add to them.

The problem with this information overload is it does not do what it sets out to do.

Like most teachers, I have to admit that I have not read every document - there just isn't time - so I have selected the parts which interest me or that I feel I ought to read and I have no control over what my colleagues select to read.

Producing more material to read probably ensures less (that we might consider important) gets read. However, recently a spokesman for the Department For Education was reported as stating: "As technology becomes easier to use, advisory services are perhaps less necessary. More material is available on how to use it and that is an effective mechanism for getting information across." (Times Educational Supplement Computers Update 10.11.92)

4 Failures and successes

If we can engender enthusiasm in teachers, they may become sufficiently motivated to try ideas and read the books. Courses for teachers should be organised to provide that stimulus by involving them in active learning themselves about something meaningful to them and where there may not be a "right" answer as in the experiment to find Young's modulus.

There has to be an acceptance that teachers do not know all the answers or their pupils will never progress to learn more than them. Teachers have to accept the fact they are attempting to make their pupils cleverer than they are themselves and that threatens their power domain.

In attempting to Promote Active Learning, I have experimented with many techniques and seen many failures and some successes.

Producing a written decree that all teachers shall work in a particular way does not work because it demoralises and demotivates skilled professionals who feel they are being criticised.
Working with schools to produce new whole school written policies does not ensure all teachers will follow them.
Enthusing teachers on courses and in workshops frequently fails when that teacher returns to the other pressures of his or her particular environment.
Working alongside the teacher in his or her classroom doesn't always ensure the active learning environment created will be sustained when the hand holding is over.

One of the most successful strategies I have employed is to actually work with children. Until recently it was fashionable to think in terms of "child-centred" education. It seems to be now "government centred" as education becomes a political weapon.
However, most teachers still respond to their pupils' demands more than governments' which is why the strategy works.

Working with the children on projects which are exciting and motivating and often left unfinished because of time considerations, means they will pester their teachers to be allowed to continue and do more. The enthusiasm will spread as children socialise at their break times and out of school so other teachers are nagged by their pupils, or even their pupils' parents, to engage in similar activity.

Summarising the common problems mitigating against the provision of the active learning environment:

1) There is a lack of teacher expertise with staff more used to a formal didactic approach where they retain control.
2) Teachers will cite lack of pupil expertise where they are concerned about their charges' reactions to a new regime which could cause disciplinary problems as the teacher hands over control, which of course does not materialise because of the pupils' new found interest in learning.
3) There is a lack of suitable resources.
4) It is difficult initially for traditional teachers to define learning outcomes using these methods until they have more experience.
5) Teachers are worried children may not achieve a perfect working model and thereby meet some degree of failure; if there is no finished product the teachers may fail to recognise the learning that has actually taken place. This is akin to wanting a "right" answer. But it is also important for children to learn to cope with "failure" if that is what it is when they recognise not everything is as easy as they initially expect.
6) Teachers do not like feeling out of control but control can inhibit learning.
7) Teachers find it difficult to identify the learning needs of their pupils and the equipment needs of the projects in this environment until they have greater expertise and experience themselves.
8) Resources are always a problem as schools are under funded. This results in utilisation of junk materials where commercial products designed to overcome unnecessary extra complications could provide better learning opportunities.

Following a six week control technology project with eleven and twelve year olds, I elicited their comments in class discussion to compare with the teachers' perceptions. These are their recorded comments:

"We were too ambitious to start with because it was the first time we had done this and most of our other work had been manageable." Perhaps teachers are prone to provide work in which it is too easy to attain a "correct" answer, to keep the children quiet, because they do not really want the problems of having to teach children?

"We had to learn to work with others who might not know as much as we do and whom we would not normally have chosen to work with."

"We didn't have enough apparatus." They listed tools and disposable materials I expected but highlighted Technic LEGO® because "it doesn't need gluing, it already has gearing, shafts and bearings provided and is recyclable".

"We needed to have learned more." There has to be a progression of skills in different disciplines. Their previous experiences with electricity had developed the concept of electric circuits requiring two wires to carry electricity "to and from" a device via a switch. Unfortunately, some control interfaces utilise single jack plugs and paired cable apparently negating that concept. (I have often seen children attempt to attach a crocodile clip to a jack plug, wondering how to complete the circuit, or ask why there is only one lead to a provided sensor.)

Unfortunately the progression had been broken the previous year and "electricity" had been dropped from their provided curriculum to make way for other aspects of National Curriculum and they felt they had had to relearn.

5 Friendly software and computers

The skills progression also calls for technological systems design and manufacturing skills to be introduced throughout the school. Unfortunately, although we are good at designing and making artefacts and environments, systems are often neglected because of the lack of suitable materials.

The third of our skills development strands is programming in LOGO. Again there have been problems with teacher confidence.

As another aid to Promoting Active Learning, I have attempted to make things easier for teachers and learned that "User friendliness" depends upon the user.
A few years ago, reacting to teacher frustration, I produced "Notepad LOGO" which could record the pupils' work on the computer and provide a simple toolkit for defining and editing procedures and printing out work. This was well received. Envisaging possible problems when introducing language uses of LOGO, I produced a series of microworlds to make it easier to use LOGO for writing adventure stories, produce electronic magazines, present quizzes etc. This was not so well received.

The new Archimedes computers we use have been designed to be "user friendly" with a "desktop" environment. To run a program, the mouse is used to drag the pointer to the disc drive icon. The mouse has three buttons on it. Clicking the left button produces a directory of the disc. The pointer then has to be moved to the appropriate program and the left button clicked twice in quick succession. A few moments later, the program icon appears on the icon bar. The pointer has to be moved to the new image and the left button clicked yet again to get the program running. Remembering which button to press, where, when and how many times was not exactly what teachers thought of as being "friendly".

I compounded the problem by providing a "front end" to the LOGO we had available for that machine so, once running, an on screen menu provided for easy selection of Notepad LOGO, Language ideas, Control LOGO etc in addition to "ordinary LOGO". It appears having to make these initial choices provides extra unnecessary complications. Teachers would prefer to be met with "ordinary LOGO" and call up the menu (possibly via a function key) if anything else is wanted.

In short, I have found no easy answers to promoting active learning but I hope my observations have produced food for thought for others.

Workshops and Discussion about Educational Situations

Martial Vivet and Pascal Leroux

Laboratoire d'Informatique Université du Maine, Avenue Olivier Messiaen - B.P. 535, F-72017 Le Mans Cedex, France

1 Introduction

This text is an abstract of the workshop managed during the meeting. For a few years our control technology activities have been focused on adults' training in technology. In a first part, with the help of hands-on, slides, videocassettes, and transparencies we have shown the different activities and tools used in such a context. A videocassette presented an experiment with "low level qualified" adults.

The second part of the workshop was dedicated to a discussion about the educational situations and particularly we faced questions like "Which robots for which abilities? Which pedagogical activities for which learners?" A robot must be adjusted to the function for which it has been built. When it has pedagogical functions, it carries different constraints and limits. The constraints and limits are not the same as those needed for an industrial robot.

2 Activities and tools

2.1 Basic notions in technology

In a first approach, the learners use micro-robots, built from Fischer-Technik® bricks and activated from a computer, to acquire basic notions in technology. This acquisition is realized from three different pedagogical activities which are: activating, building from diagrams or designing micro-robots.

Two workshops were organised to show the learner's work environment. Attendees had the possibility to activate and play with an arm and a CARISTO. The CARISTO is a first robot to reach the third dimension. Directly derived from the floor turtle by Papert, the shell of the turtle is only changed into a coachbuilding fork-lift truck, the pen being changed into a fork.

We organized a presentation of slides and videocassettes showing engines sorting French coins or sorting, storing at different stairs with a lift, cubes according to their colour. These engines had been designed by adults from specifications and functions given by the teacher.

2.2 Automation and flexibility concepts

In a second approach, in order to learn automation and flexibility concepts in production processing factories, the learners use pre-constructed tools, the AFX and the SPF. The goal is not to train student on a specific machine or in specific ways to describe an automatic process of production. The goal is here to train them to make decisions and to manage a production process from a customer to its delivery.

The SPF (System of Flexible Production) is a tool to learn flexibility concepts in production. This environment occupies a complete large room and allows an effective production controlled by professional software (TIMELINE) to approach the problems of Computer Integrated Manufacturing (CIM).

The AFX (Atelier FleXible) is a Fischer-Technik® machine standing on a table (1 square meter). The approach looks like the one with SPF but the tool is less expensive. There is no danger when manipulating it, so failure can occur. We have now defined pedagogical scenarios to use AFX. Some allow help-to-learn reading screens which represent a production process. An important work has been done on this approach of apprehending control of a physical world via screen representations. With low qualified adults, the main problem is acquiring confidence with what is written on the screen with respect to what occurs in the real world. Another problem is to train them reading the right thing at the right moment and interpreting what is written, getting sense of the meaning; and finally gaining a correct answer when interpreting an event into real world with the only information written on the screen. Software to manage a production is now available.

3 Videocassette of an experiment (QUADRATURE) with "low level qualified" adults

QUADRATURE is a project funded by the French Ministery for Research and Technologies. This project used modular micro-robots to help "low level qualified" adults acquiring new skills. This experiment has been managed by a synergy of different partners: a manufacturing company, a labour training association and the computer science laboratory (LIUM). First results have shown that the tools are relevant for vocational reconversion operators so that they will be able to work on an automated production line. Through the handling of these micro-robots, they will acquire basic knowledge and abilities.

References

Vivet, M., Bruneau, J., Parmentier, C. (1990) Learning with micro-robotics activities. NATO Advanced Research Workshop, Eindoven, October 9/12 1990. In: M. Hacker; A. Gordon, M. de Vries (eds.), Integrating Advanced Technology into Technology Education. NATO ASI Series F, Vol. 78, pp. 139-148. Berlin: Springer-Verlag, 1991.

Vivet, M., Parmentier, C. (1991) Low qualified adults in computer integrated enterprise: an example of in service training. IFIP TC3/WG3.4, Alesund, Norway, 1-5 July 1991. In B.Z. Barta and H. Haugen (eds.): TRAINING: from Computer Aided Design to Computer Integrated Enterprise, pp. 261-272, North-Holland, 1991.

2. Experiments and Case Studies

2.3 The Teacher's Role

Educational Uses of Control Technology

Martial Vivet

Laboratoire informatique, Université du Maine, Avenue Olivier Messiaen
BP 535, F-72017 Le Mans Cedex, France

Abstract. The text discusses project based teaching. We are working mainly on the use of micro-robots to train people with relevant basic knowledge around robotics and computer integrated manufacturing (CIM). First the text clarifies what is a project in our approach. In short, a project has a pedagogical goal in terms of contents, abilities and attitudes to be learned. The approach is based on a contract between a teacher and a group a learners to produce an effective machine. We describe then the main characteristics of "good projects". The focus is on the role and the attitude of the teacher seen as a manager of a learning environment. Different phases in the work and the learning process are identified: negotiation of a contract, management of a contract, reception/validation of the production. The problem of evaluation is faced and linked to the collective evaluation of different technical solutions to a given problem. We underline there the change in the role of the teacher who no longer manages tools, sheets, or bits of kit, but manages a very fruitful discussion.

An example of such a project is given in which the described approach is used in order to retrain lowly skilled adults.

Keywords. Adult learning, Control technology, Cooperative learning, Pedagogical attitude, Project driven learning, Pedagogical robotics, Technology learning.

1 Introduction

Building learning environments for technology is a tremendous educational challenge. The main difficulty come from a very often implicit goal that consists in giving people the mental attitude of engineers (able to use effective knowledge to build solution to practical problems) and the mental attitude of researchers (able to create new knowledge by themselves, leaving the usual conception of learning based only on knowledge transmission). We propose a project based approach to train learners acquiring such abilities. At the pedagogical level, we try to use contrainsts like those imposed every day to engineers or researchers. This allows people to be effective while working in groups. We have used this pedagogical approach to train lowly skilled adults to be reconverted from traditional factories to Computer Integrated Manufactories (CIM). In earlier texts (Vivet 1990, 1991,

Parmentier 1990, 1991) we have mainly described the experiments. We try here to reach a better pedagogical formalisation of our action.

2 Project Based Teaching

2.1 What is a Project?

Our approaches to teach technology are mainly based on projects whose realisation is conducted by the learners. Learners act as a team and the teacher appears as a project manager as usualy understood in industry. A project is defined by its goals, its content, its constraints (costs and time). Project we are working on deal with a cultural approach of Computer Integrated Manufacturing (CIM), computer integrated engineering (CIE). They are mainly based on design, construction and control of micro-robots.

2.1.1 Goals

For a given project, we define two kinds of goals; the *pedagogical goal* is followed by the teacher who has the responsability to manage the learning process, the learning situations; the *technological goal* deals with the technical device which is to be built and controlled by the learners. Even if they can be strongly connected by the fact that the learning process will be organised, facilitated by the self involvement of the learner to reach the technical goal, pedagogical and technical goals must be clearly separated.
- *pedagogical goal*: For each project, the pedagogical goal must be clarified in terms of contents and attitudes to be learned. The contents can refer to technical knowledge, know-how but also to meta-cognitive skills and mental abilities.
- *technological goal*: generally refers to the production and the control of a working device. As far as possible, technical functions of such devices are "similar" to functions effectively in use in process control in industry. The main problem to be faced is the problem of transfer: what can be transfered (and how to help such a transfer?) from the learning situation to the work place? The "similarity" function would need to be clarified in each case.

2.1.2 A Content

The technological content of the project must be carefully chosen involving fundamental knowledge in the domain, habilities to be learnt and know how to be acquired. It is very useful to describe at the early beginning the kind of content we want the learner to master.

For example, working on kinematics chains impose projects where you can have an incremental design with at first only one kinematic chain. Later without leaving coherence, without losing too much time completing the devices, it can be interesting to reach a stage where it is possible to compose kinematic chains.

A possible pedagogical approach is to work with learners organised in small groups, each group working on a device offering a given kinematic chain. The coherence must be planified to allow in a later stage cooperation between machines, each group offering its own machine to build a more complex one. Several advantages are met that way : motivation can then be very high, it is possible to share the work, mapping the complexity of each device according to competences available in each group, in the final sequence. Doing so, the collective success of the project is the success of everybody (even for the learner encountering more difficulties). From the project management's point of view, its is clear that obligation of success in a given time for each sub-project is a very motivating constraint.

Few necessary, even not sufficient, characteristics to reach success will be developed hereafter.

2.1.3 Successful Characteristics

Motivation. As far as possible, the similarity with real world must be evident for the learner. That means that the learner must feel confortable with the technological goal and the content. This can be obtained if the device to be designed offers functions easily understood. We have worked for example on sorting machines (e.g. sorting coins according to their values or cubes according to their sizes or colors). Motivation conditions can be better met by starting with devices "looking like" those met in every day life. When retraining adult workers, starting with robots similar (from geometrical, mechanical point of views) to those used in the factory is a very strong help. With younger pupils, copying robots seen during the visit in a factory can help. A possible organisation is to make at first drawings and paintings of these robots and later to build and drive reductions of similar ones. We can also find articulation of arts and technology. Works done in Québec (Lecompte 1990) around control of carroussel or animated Christmas shops are really very characteristic of these ideas.

No evident solution. A project is better if there is no evident solution even for the teacher who never solved it. We discussed earlier (Vivet 1991) the difference for us between exercises and problem solving. It is very important to have a teacher playing the role of a researcher in such environments. So if the teacher him(her)self does not have an evident solution, (s)he will not impose it and this is very important for the learners.

Different solutions. A project is far better if different technological solutions can be found for the project. This allows the learner to find better solutions than those prepared by other groups or by the teacher. This can be a very motivating challenge because it creates conditions for real technical argumentation to justify choices. Such stages can be very important to re-enforce meta-cognitive skills. About the pedagogical organisation of such a work, this means that it is very important to manage time for these stages such as time for sharing solutions, comparing, justifying them. For us, these stages are managed during the reception phase of the devices.

Target device allowing different point of view. The target device is complex, allows apparent useful functions. The need for project whose technological goal is to produce devices allowing different points of view appears as very important to us and, therefore, during the learning sessions, we manage time to discuss these points of views. The learners are very motivated by this confrontation of their solution with the solutions produced by others (other groups' solutions or the teacher's one!).

As in industrial projects, a very important phase is the reception time. During this phase you must justify that predefined constrainsts are satisfied and the designer/producer of the device must show the advantages of the offered solution.

Criteria to judge the productions can be classified into :
- technical criteria of the device : fiability, maintenance, evolutivity, portability, interoperability, re-usability of modules, etc
- economics : analysis of costs (costs of the machine alone (at delivery time!), costs of the machine while working (number of workers necessary to use the machinery (definition of each work place)), productivity (number of pieces produced in one hour! -at least - at best - in average according to failures), cost in case of failure (MTBF : concept of Mean Time Between Failure), time of stop, mean time for maintenance,...
- security : protection of workers around the machine while working, protection during maintenance activities,...
- usability : ergonomics at the work place for workers using the machine, role of the workers.

The definition of such criteria gives a possible plan to manage the discussion with the learners. Such a discussion can be very long but appears to be very useful. At its end, the teacher can show his/her own solution or solutions produced in earlier groups. Sometimes, the challenge is won by the learners who found "better" solutions than thoses proposed that way by the teacher.

It is clear than this can affect deeply (and positively) the human relations between the learners (among themselves) and the teacher.

Work in different sub groups. The kind of pedagogical dynamics we just described is possible by chosing projects allowing different solutions which can be produced separately by different sub-groups. Each sub-group works on the same technical goal. This can allow failure for some groups unable to finish in a given time without destruction of the work for the whole group. This approach can be chosen with classes with high degree of heterogeneity (i.e. very different levels between the learners). The lowest groups can be completely engaged in the final discussion discovering by themselves then why they failed.

Projects splitted in sub-projects where subtasks needing "similar effort" to solve them can be also very interesting. This can be used if the lowest sub-group is not bad enough to impose failure for the whole project. But reserving to such a group an easy sub-task can be a solution. More: a longer time can be spent by the teacher with such a sub-group. The projects and tasks being different in each sub-group can be an acceptable reason for the teacher not to spend the same time with each sub-group.

So according to the whole set of learners the teachers have to work with, the kind of contract must be adapted.

2.2 Pedagogical Attitudes

2.2.1 Building on Implicit Knowledge

The first step in the design process for a new project is to define and clarify the pedagogical goals.Then the design of the content begins leading to a technological content. It is very important to facilitate transfers by building on implicit knowledge. So the technological goals of the device to be built must refer to already available even if not conscious knowledge. For example, with adults working every day in a manufacturing factory, starting with robots looking like thoses available and seen in the factory can be very important. This allows a faster first approach with reduced functions. Later on, discussions appear because of perception of differences ("Eh! it is not like this in the factory. There, the process stops only when..."). You can move on to include new functionalities. You have here a nice way to invite people to examine the place where they live every day, to understand machines and process around them. They get a different view of their workplace.

This helps you managing your didactical situation! This kind of approach imposes a very good understanding of the factory where the learners come from. So preparing the next "course" more or less consists in visiting the factory.

With children the approach can be similar even if different. A visit in a factory, followed by drawing/painting or a presentation of a video / TV can help starting the process. Very good lessons to learn about technical observation can be managed. Coming back to the factory at the end of the project can underline the differences and reinforce the learned process.

2.2.2 Elicitation of Knowledge

A main aspect in such approaches is to leave the work only based on implicit knowledge to reach an effective elicitation of knowledge. This means pushing the learner from working with his/her hands, by intuition, to a more explicited or formalised way. The need for communication among groups can help formalisation at the level of language. One of the difficulties is to manage time, allowing the use of the learner's language for "just long enough time". The teacher must introduce the usual and conventional language at the right time: after the concept is understood and before intensive use of words to communicate with others. An effective help to fix the right language can come from sheets giving helps and additional information. Here, hypermedia tools can probably play an effective role.

2.2.3 Working in Groups

In our approach, most of the work is done in small groups of learners. The composition of the groups must be checked carefully. According to the context, it is possible to mix or separate male and female participants, merge people with a good background on the factory process with novice (virgin) people with respect to this technical culture. The decision is applied taking account of the possibility to split the project into very similar or very different subtasks. The required implicit knowledge and competences useful to solve the problems in such sub-tasks can be different and here is the place to manage the heterogeneity of the group.

It is sometimes possible to split the available competences into different types of groups which become specialised on different functions; for example, it is possible to organise groups for production, groups for management (management of a store of available modules, management of costs,...). We can observe the competition between the "producers groups" needing the same engine, the last available one on the shelves! How do the groups solve such a problem? This can be solved by social interaction (a fight outside during a break?), a change in a technical solution (who really needs this engine and is unable to solve the local problem without it?). We can find times to have groups discussing solutions produced by others, helping to find different solutions; These are very good opportunities to learn, to revisit one's own solutions. Indeed, as teachers, we must appreciate the moment when the last engine is to be used! The management of penury in education can sometimes be useful to help the learning process!

In each project, the teacher must manage carefully a contract with each group, the constraints being justified from constraints coming from other groups. Managing time for social interaction between groups is here important and "losing time" at the beginning of a project can be a very effective gain for later developments!

2.2.4 The Teacher's Attitudes

With such approaches, the teacher plays different roles such as:
- Architect: In each project, the teacher works above all as a designer, an architect organising a "living place, a learning space". His (her) works is at first to design the learning context according to the environment e.g. who are the learners? (each of them having a useful competence to be used), appreciation of heterogeneity, organisation of tasks and activities, preparation of ressources)
- contract manager: contract for a device to be produced according to constraints accepted by everybody, each sub-group having to face constraints coming from other groups, from time, etc.
- animator: While the project is going on, the teacher's main role consists in giving advises and helps. It can help in management of resources (this last role can be different if a specific sub-group is in charge of this task).
  A hiden role is to accept and manage confidences made by each group which can wish to put trust in the teacher. Confidences made can concern the technical solutions, or the need for specific helps, etc.
- reception of furniture: The role is here that of a contractor receiving a product according to specifications. The role is mainly to organise a discussion converging in a comparison of the different devices which are built. Generally each given solution presents enough positive aspects to have everybody feeling successful somewhere. Appreciation of deliveries can be done according to the different points of views described in 2.1.3. A possible plan to manage the discussion follows clearly the given list of criteria. For this aspect of

evaluation, it is valuable to reach co-evaluation by learners during the discussion. This is a good time to come back again on what has been worked, discovering solutions by others, justifying one's own solutions even accepting that the others' ones are better. The role and importance of such phases at the metacognitive levels must be underlined once more.

2.3 Evaluation

We must separate clearly between evaluation of the production as done during the reception of deliveries and the evaluation of what has been learned during such a project. Mainly during such projects, the learners learn how to manage work with others, taking account of time or technical constrainsts. They also learn how to be effective. Mainly based on their personnal activity, they learn the attitude of engineers. From the technological content, they learn how to use and exploit technical documents, how to describe technical devices. But the need for more formalised ways of evaluation is still here. The compatibility with examinations as usually done is not sure! So how could the results be measured?

3 Conclusion

This text is an attempt to reach a pedagogical formulation of project based teaching. Such an approach appeared during several training sessions in schools or in industry as a very effective one to help people learning technology not only in its content but also in the way to live it. The kind of integration of courses, theory and practice we reach by this way is interesting. In any case the very strong motivation the learners show for such activities encourages us. Among remaining problems is the hyper-activity of the teacher while managing such groups. We [LEROUX 91&92] are working to design "intelligent" helps for such teachers.

Acknowledgement

Thanks are due to the French Ministry of Research and Technology (MRT), Glaenzer-Spicer, Scitec companies and the training center of the "Association pour la Formation Professionnelle" (AFP) in Le Mans, who jointly supported research allowing evolution of such ideas. Projects like QUADRATURE, PLUME, PALOURDE are concerned.

References

Baudry, M.: "SPF : un environnement d'apprentissage des concepts liés à la flexibilité dans la production", Premier congrès francophone de robotique pédagogique, Le Mans, 30/8-1/9, 1989.

Bruneau, J.: "Remarques autour d'une activité de robotique en classe de 6ième", Premier congrès francophone de robotique pédagogique, Le Mans, 30/8-1/9, 1989.

Charlot, B.: L'école en mutation: Payot, Paris, 1987.

Dickson, P.: "Environment for interactive learning; the computer exploratorium", NATO ASI, Calgary, July 16-28, 1990. In: M. Jones, P.H. Winne (eds.), Adaptive Learning Environments. NATO ASI Series F, Vol. 85. Berlin, Springer-Verlag, 1992.

Enrique Ruiz-Velasco Sanchez: "Un robot pédagogique pour l'apprentissage de concepts informatiques", Premier congrès francophone de robotique pédagogique, Le Mans, 30/8-1/9, 1989.

Giovannini, M.L.: "Robotique pédagogique à l'école primaire ; Une première expérimentation Légo-TC-LOGO à Bologne", Premier congrès francophone de robotique pédagogique, Le Mans, 30/8-1/9, 1989.

Gordon, A.: "New technology in secondary schools: problem solving", NATO ARW, Milton Keynes, November 11-12, 1988. In: E. Scanlon, T. O'Shea (eds.), New Directions in Educational Technology, NATO ASI Series F, Vol. 96. Berlin: Springer-Verlag, 1992.

Lecompte L., Moquin F.: "Gestion d'un projet de robotique pédagogique au premier cycle du secondaire", Deuxième congrès francophone de robotique pédagogique, Montréal, 26-28 August, 1990, pp.111-118.

Leroux, P., Bruneau J.: "Coopération entre un élève, un environnement de micro-robotique et un système expert de pilotage de micro-robots", 3ème congrès robotique pédagogique, Mexico, August 1991.

Leroux, P.: "Cooperation between pupil and expert system to drive a micro-robot", ICTE, Weimar, 25-30 April 1992.

Leroux, P., Bruneau, J.: "Activating a micro-robot without using a programming language", ICCAL 92, Wolfville, June 1992, pp. 48-50.

Marchand, D.: "Construction et programmation par des élèves d'un micro-robot", Premier congrès francophone de robotique pédagogique, Le Mans, 30/8-1/9, 1989.

Mercier, J.: "Modélisation du processus d'apprentissage des objets techniques", Premier congrès francophone de robotique pédagogique, Le Mans, 30/8-1/9, 1989.

Nonnon, P : "La robotique pédagogique", revue BUS, AQUOPS Montréal, May 1987, pp.16-19.

Nonnon, P., Laurencelle, L.: "L'appariteur robot et la pédagogie des disciplines expérimentales", SPECTRE, pp. 34-36.

Papert, S.: Mindstorms: children, computers and powerful ideas. Basic Books, New York.

Parmentier, C., Vivet, M.: "Micro-robots et QUADRATURE : qualification, demande en reconversion d'adultes et technologie utilisant des robots éducatifs", actes de l'université d'été, ISHA "Informatique & apprentissages", Chatenay Malabry, Octobre 1990. Paris, l'INRP, 1991, pp 91-106.

Parmentier, C., Vivet, M., Bruneau, J.: "Micro-robots: object or tool for factory training?", poster CAL91, Lancaster, 8-11 April, 1991.

Parmentier, C., Vivet, M., Bruneau, J.: "La reconversion d'ouvriers spécialisés: un défi pour la reconversion d'ouvriers spécialisés". Paru dans les Actes du Second congrès francophone de robotique pédagogique. Montréal, August 1990.

Parmentier, C., Vivet, M., Bruneau, J.: "LOGO et la reconversion d'ouvriers", Colloque LOGO & Apprentissages, Fribourg, October 1990.

Parmentier, C., Vivet, M., Bruneau, J.: "Robotique objet/outil pour la reconversion d'Ouvriers Spécialisés", Congrès APPLICA, Lille, Sept. 1990.

Pelchat, R.: "Le robot enfouisseur de déchets nucléaires, Guide d'activités", CARES, 10450 rue Meunier, Montréal, H3L 2Z4, Québec.

Rabardel, P.& Verillon, P.: "Robotique pédagogique et conceptualisation du repérage tri-dimensionnel dans l'espace", Premier congrès francophone de robotique pédagogique, Le Mans, 30/8-1/9, 1989.

Schoefs, Y.: "Exploitation pédagogique du SPF: premier bilan d'une expérimentation", Premier congrès francophone de robotique pédagogique, Le Mans, 30/8-1/9, 1989.

Tanguy, R.: "Un réseau de mobiles autonomes pour l'apprentissage de la communication", thèse d'université, Paris VI. 1987.

Thornton, R.: "Using technology in the construction of scientific knowledge", NATO ARW, Milton Keynes, November 11-12, 1988. In: E. Scanlon, T. O'Shea (eds.), New Directions in Educational Technology, NATO ASI Series F, Vol. 96. Berlin: Springer-Verlag, 1992.

Vergnaud, G.: "Concepts et schèmes dans une théorie opératoire de la représentation", Psychologie Française, n°30-3/4, pp. 245-252.

Vivet, M: "Learning Science & engineering with open knowledge based systems", in E. Forte (Ed.) CALISCE '91, September 9-11, 1991, Lausanne: Presses polytechniques et universitaires Romandes, pp. 53-64.

Vivet, M.: "Knowledge based systems for education: taking in account the learner's context", PEG-91, Rappallo, in Computers & Education (Pergamon Press) Special issue, May 1991.

Vivet, M.: "Uses of ITS, which role for the teacher?", NATO ARW, Sintra, October 6/10, 1990. In: E. Costa (ed.), New Directions for Intelligent Tutoring Systems. NATO ASI Series F, Vol. 91. Berlin: Springer-Verlag, 1992.

Vivet, M., Bruneau, J., Parmentier, C.: "Learning with micro-robotics activities", NATO Advanced Research Workshop: Eindhoven, October 10/12, 1990. In: M. Hacker, A. Gordon, M. de Vries (eds.), Integrating Advanced Technology into Technology Education. NATO ASI Series F, Vol. 78. Berlin: Springer-Verlag, 1991.

Vivet, M., "Démarches de modélisation en robotique pédagogique", 3ème congrès robotique pédagogique, Mexico, August 1991.

Vivet, M., Parmentier, C.: "Low qualified adults in computer integrated enterprise : an example of in service training", IFIP /TC3 meeting, Alesund (Norway), 1-5 July 1991. North Holland, 1991, Eds B. Z. Barta & H. Haugen, pp. 261-272.

Vivet M.: "Apprentissage autonome, sur un usage de la technologie informatique dans l'éducation". Annexe du rapport SIMON, "l'éducation et l'Informatisation de la Société". La documentation française. pp. 201-210.

Vivet, M.: "Driving Micro-robots under LOGO : a way to approach geometry". Proceedings of the second international conference for LOGO and MATHS education. Institute of education. University of London, 15/18 July 1986, pp. 216-225.

Vivet, M.: "Examination of two ways for research in advanced educational technology", NATO ARW, Milton Keynes, November 11-12, 1988. In: E. Scanlon, T. O'Shea (eds.), New Directions in Educational Technology, NATO ASI Series F, Vol. 96. Berlin: Springer-Verlag, 1992.

Vivet, M.: "LOGO: un outil pour une formation de base à la robotique". Colloque National "LOGO et les enseignements technologiques" - Le Mans - Nov. 83 , compte-rendu revue ETI-LISH/CNRS, n° 3, pp.5-19, 1983.

Vivet, M.: "Micro-robots as a source of motivation for geometry", Advanced research workshop "modelling the student knowledge, the case of geometry", NATO ARW, Grenoble, November 13/16, 1989. In: J.M. Laborde (ed.), Intelligent Learning Environments: The Case of Geometry. NATO ASI Series F, Vol. 117. Berlin: Springer-Verlag, 1993.

Vivet, M.: "Pilotage de Micro-robots sous LOGO, un outil pour sensibiliser les personnels de l'industrie à la robotique" - 5e symposium canadien sur la technologie pédagogique - Ottawa - 5/7 mai 1986. Paru dans le livre "A l'école des robots" , robothèque du CESTA, pp. 195-210.

Vivet, M.: "Robotique pédagogique! soit, mais pour enseigner quoi?", Premier congrès francophone de robotique pédagogique, Le Mans, 30/8-1/9, 1989.

Vivet, M. : "Which goals, with which pedagogical attitudes with micro-robots in a classroom?", NATO ARW on Advanced Educational Technology, Pavia, Italy, 4/7 October 1989

Vivet, M., Parmentier, C, Bruneau, J.: "La reconversion d'ouvriers spécialisés : un défi pour la robotique pédagogique", Deuxième congrès francophone de robotique pédagogique, Montréal, 26-28 August, 1990.

Measuring some Cognitive Effects of Using Control Technology

Brigitte Denis

Service de Technologie de l'Education, Université de Liège au Sart-Tilman, Bât. B32, B-4000 Liège, Belgium. Tel.: 32-41-56.20.72, Fax: 32-41-56.29.53, E-mail: U017801 AT BLIULG11

Abstract. Different approaches can be used to evaluate the effects of using microworlds such as control technology. Our approach is based on the regulation process and the use of established tools to observe animator and user behaviour. Some user objectives are established as well as a theoretical animator profile which could promote their attainment. Users' objectives deal with computing and technological literacy, socialization, problem solving and individual cognitive development. Observation grids have also been developed to evaluate whether there is a gap between intentions or claims and action and whether the animators' evaluation of their activity is realistic. The observation of users' and animators' behaviours in a 5th grade primary classroom provided some practical information to help the animators to take regulation decisions and then to increase their congruence and to reach their objectives.

Keywords. Assessment, Constructivism, Control technology, Elementary education, Evaluation, LOGO, Pedagogical robotics, Primary school, Problem solving, Project driven learning, Pupil learning, Observation, Regulation, Socialisation.

1 Introduction

Training needs are rapidly changing in our Occidental culture. Economic and social needs in competent, autonomous and creative citizens are increasing so much that training systems are questioned (De Landsheere, 1989; Leclercq, 1992).

To take up this educational challenge, different microworlds have been developed, e.g. LOGO microworlds, control technology, ... These microworlds have been considered by some educationists as a panacea that could solve the future educational problems.

A lot of educational objectives have been assigned to the use of these microworlds: develop learner's socialization, creativity, new intellectual structures, motivation to learn, ...

After the messianic period the evaluation period came. The debate about effects was not limited to statements and theoretical positions. Some researchers tried to make experimental studies on the impact of these microworlds on cognitive, affective and social users' development. Therefore a lot of "results" had been collected. But, often, these researches did not take the complexity of the educational process into account.

One of the dimension of this complexity concerns the *objectives* the animator chooses and evaluates. In this respect results of experimental researches are far from being generalizable since they consider a lot of different target goals : planning competency, acquisition of specific notions, creativity, ...

A second dimension concerns the *target populations*. These populations have very different characteristics: age, sex, activity experience, ... often, individual characteristics are ignored.

A third dimension concerns the *methodology of animation*. This can be very different (constructivist, directive, non directive) from one research to another. Nevertheless, it is not the microworld itself that induces positive (or negative) effects on learners' behaviours, but the actual interactions with the microworld where the animator is acting as a cognitive and social catalysor.

The context as well as the resources can consider additional dimensions affecting the results of microworld animations.

2 Methods of Evaluation

Different approaches can be used to evaluate the effects of acting in a microworld. Firstly, the "products" (dependent variable) may be evaluated. This kind of study is often a correlative one : a study of the links between learners' results to a test and the work done on the topic. Interactions between learners and animators are not studied. Nevertheless, Bloom (1979) stressed how important are actual (not theoretical) learning opportunities.

A second approach links *educational process* and *products* (independent variable + dependent variable). Researchers try to understand the knowledge building process of the learner. These studies are often case studies. Tools to help to interpret interactions between learners and animators are scarce.

A third approach deals with *regulation,* where the animation process is the target study object and is the dependent variable of a meta process (an analysis of the process). This analysis helps us to understand our practice better, to evaluate its results, and to adapt it to target goals.

3 Regulation Process

Our approach is based on the regulation process and on the use of tools to observe animator's and user's behaviours during the activity. We have distinguished five phases in the regulation process (Leclercq, 1976). They can be considered from the point of view of the animator and from that of the user respectively.

The phases are:
1. The definition of the epistemological approach of teaching and the definition of the general objectives of learning.
2. The definition of the operational objectives and of the methodology (target user's and animator's profiles) to be implemented during the activity.

3. The observation of animator's and users' behaviours.
4. The measurement of effects.
5. Regulation decisions.

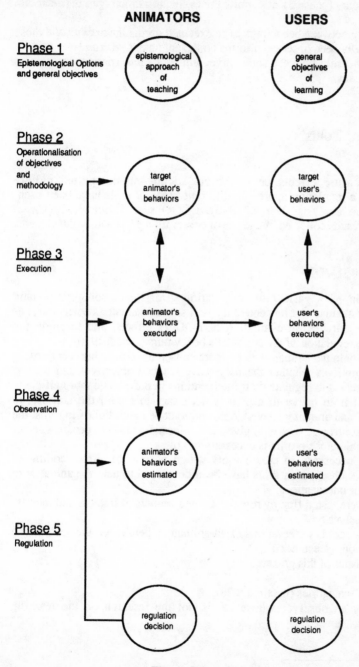

Fig. 1. Regulation Process

The regulation process helps in the study of the animator's congruence, realism and efficiency.

Congruence is the study of the links between theoretical profiles (phase 2) and observed behaviours (phase 3). It permits the evaluation of any gap between talk and action.

The comparison between behaviours to be executed during the activity and those which have actually been observed informs on the animator's *realism* [1].

The study of the links between some target animator's interventions informs on animator's *efficiency*.

4 Evaluation Tools

In order to measure effects on cognitive, social and computing culture development in a regulation approach, different kinds of tools have been used: grids, sequential analysis (Denis, 1990), LOGO scan (Sougné, 1990), ... Hereafter, only results collected by the use of observation grids will be developed.

4.1 Observation Grids

We have developed a variety of observation grids. One, containing many categories (155) against which is coded the user's or an animator's behaviour, has been used in researches on evaluation of LOGO's effects; and to study the feasibility of the regulation of the animator's behaviours (Denis, 1990).

Our goal is to help the animators to be more conscious of their interventions, to evaluate by themselves whether the users have reached objectives and then to decide by themselves to regulate their interventions in order to be more efficient. But such a grid is to heavy to use: it is necessary to record the activity, to transcribe users' and animators' behaviours and to code them. Following this, the analysis of the results was needed to give some information to help decide whether a regulation of animator's profile is necessary or not.

Thus, it was necessary to create tools which permit the direct coding of interactions. Two observation grids have been developed to help the animator to self-regulate their interventions.

Those grids were tested first by researchers and trainers of trainers and later by animators themselves.

The first grid records observations of an animator's behaviour, the second one the users' behaviour (see annex 1).

The characteristics of this grid are:
- direct coding,
- "synthetic" categories (maximum 28),
- usable by an observer whose job is not the research on the learning processes,

---

[1] For further developments about realism, see Leclercq (1992b).

- usable by the animator him(her)self or by an observer,
- immediate interpretation of results.

Our hypothesis was that training the animators in the use of this kind of grid could help them to increase their congruence and their realism in relation to the objectives they wanted to reach and to be autonomous in making decisions on the regulation of their own behaviour.

4.1.1 The Best Tools?

Solutions other than the appropriation of a tool developed by researchers could have been considered (e.g. development of observation grids by animators themselves or capturing information on the activity by other recording media, interviews of students, ...). But the tools presented here seem to provide an economical solution leading the animators to be more objective about what actually happens during an activity.

4.1.2 Is Direct Self-observation Feasible?

If we consider an un-aided regulation of the activity by the animator, only the users' grid may be used. Clearly, it is not possible to self-observe and code ones own behaviour during interactions with users. However, some information, such as the number of interventions with users and their spontaneity, would not be difficult for the animator to record him(her)self.

The animators have to take into account a very complex process. They cannot predict with certainty a learner's behaviour, they have to catch opportunities of intervening and manage an environment where individualization and active learning are basic principles. Asking them to code and interpret results is imposing an additional task with which they are not familiar.

4.1.3 Observation Between Animators?

Whilst establishing self-regulation of his (her) behaviour, the animator can be helped by an observer (e.g. an other animator) who collects information on his(her) profile or on the user's behaviours.

4.1.4 Description of Observation Grids

General structure. These grids are called "variable geometry grids" since it is possible to code a behaviour at different levels. Each category is divided into sub-categories which take into account more and more specific information on the target behaviour.

For instance, inside the category "action" of the users' grid, we can specify whether the user is acting on the computer, on control technology, on other hardware, on a reference book, etc.

In the same way, the category "verbalization" is divided into elicitations, evaluation, ... which can be further specified after (e.g. elicit an explanation on how to solve a problem, elicit a prediction on the effect of the action, ...).

Of course, these grids are focussed on specific target behaviour concerning the animator's and the user's objectives. These objectives are detailed hereafter.

The animator's grid. In addition to the categories in the grid, other information is collected to answer different questions:
- Is the animator's intervention spontaneous or not ?
- How many sequences of intervention concern different users ?
- Who is calling the most often on the animator?
- What is the average length of animator's sequences of intervention?

The users' grid. Certain information is collected to help to interpret the results:
- Who has been observed ?
- By whom ?
- How long was the observation ?
- What is the context of the activity (how many computers ? How many users ? ...).

4.2 Sequential analysis

Sequential analysis of a target behaviour is focussed on *antecedents* (what happens before) and *consequents* (what happens after).
We study whether the target behaviour has had the expected effect; and which contexts (antecedents) favour *the efficiency* of attainment of the target behaviour.

4.3 LOGO SCAN

LOGO Scan is a multidimensional package developed by Sougné (1990) to analyse the structure, the chunking optimisation and the contents of users' LOGO procedures.

The *structure analysis* of the user's programs indicates their general structure giving its organigram. Following this, we can study the organization of the different sub-procedures. The chunking evaluates the internal structuration of each procedure.

The *content analysis* gives an inventory of the primitives in the user's program. With this inventory, we can analyse the development of each user's mastery of LOGO.

5 User's Target Objectives

The objectives of a control technology activity can be numerous. We present here a list assembled with the agreement of the animators who have been observed. These objectives do not deal with a specific content such as electricity, mechanics, etc., but with the development of
- computing literacy,
- socialization,
- problem solving process,
- individual opening out.

Each of these objectives is illustrated by three more specific aaspects of behaviour the user should develop.

5.1 Computing literacy

- Demystify the computer.
- Discover and use basic computing notions (e.g. procedure, variable, ...) and construct them in a functional context;
- Develop a critical and active approach to the computer environment.

5.2 Socialization

- Increase social interactions such as socio-cognitive conflicts, cooperation and communication.
- Develop verbalization by explaining his (her) actions, his (her) strategies.
- Adapt him(her)self to different social situations; take part in a group for joint activities.

5.3 Problem solving process

- Formulate and categorize his (her) problems and divide them up to sub-problems.
- Formulate hypotheses about possible solutions.
- Improve hypotheses and evaluate results.

5.4 Individual opening out

- Increase self confidence by adopting a positive attitude towards errors and mistakes.
- Develop creativity and innovation thinking through the formulation and realization of projects.
- Develop autonomy and take care of his (her) own learning.

6 The User's Target Profile

During the current research, we specifically focussed our observation on the following user behaviour sequence:

* Conceive the project.
  - State the effect which is going to be produced.
  - Evaluate his(her) idea or his(her) project.

* Prepare the realization of the problem.
  - Consult reference documents.
  - Write notes.
  - Describe the action needed to produce an effect.

* Realize his (her) project.
  - Use LOGO instructions correctly .
  - Predict the result of his (her) actions.
  - Describe his (her) actions.

* Analyse the results of his (her) action.
  - Describe the reaction of the objects to his (her) action.
  - Describe the dependency relation between the effect and the action.
  - Describe or show the action intended to correct or carry out this one.

7 The Animator's Target Objectives

We also defined a list of objectives that we consider an animator should reach. These general objectives are linked to the user's ones. We anticipate that if the animator is trained in this way, he (she) would be both congruent and efficient.

These objectives are related to:
  - computing training;
  - social training;
  - training in constructivism;
  - practice in formative evaluation.

Each is specified by three operational objectives (behaviours) a LOGO animator should exhibit.

7.1 Computing training

This one leads him (her) to:
  - know how a computer works and how to manipulate it.
  - choose materials and tools (computer language, ...) adapted to his (her) own problems.
  - use these choices to solve problems efficiently.

7.2 Social training

This one leads him (her) to:
  - work with others, organize his (her) work in cooperation with others.
  - apply fundamental dynamic group principles related to the management of learning environments.
  - know the impact of different kinds of social interaction which have an influence on cognitive development, such as socio-cognitive conflicts, cooperation, ...

7.3 Training in constructivism

This one leads him (her) to:
  - develop active methodology based on the user's project realization, elicitate efficiently effects to produce, encourage experiences, initiatives, creativity, ...

- develop structured problem solving strategies, verbalizations about dependence into actions and produced effect, about predictions, about anticipation, ...
- elicitate organization of knowledge, collection of information and research into objects and persons in problem solving.

7.4 Practice in formative evaluation

This one leads him (her) to:
- auto-evaluate continuously his teaching behaviour and understand user's knowledge building processes.
- regulate his(her) interventions to be as efficient as possible, choose the right time to teach in an individual or collective mode.
- exploit and reveal the maximum potential of each individual.

8 The Animator's Target Profile

Discussion with the animators who worked with us led us to establish an animator's target profile. It is expected to enhance the emergence of the target user's behaviours.

The theoretical reference profile postulates that the animator should:

1. Develop elicitations about:
   - invitation to continue action.
   - effects to be produced.
   - variation of actions and effects.
   - verbalizations such as:
   - description of project or future action.
   - description of the dependence relation between a produced effect and an action.
   - description of an expected result.
   - description of an object reaction.
   - description of an action.

2. Favour:
   - positive evaluations of action and project.
   - encouragements.
   - positive descriptions.
   - planning, projects organization.
   - keeping notes.
   - referring to guides.
   - reading information given on the screen.
   - cooperation and decentration on another user's action.

This outline does not claim to be an exhaustive list of all the behaviour an animator should adopt in his (her) classroom. A teacher does not always manifest behaviour which directly relate to learning.

For instance, organizational and classroom management behaviour is more frequent than developmental and personalized behaviours in a traditional classroom activity (G. De Landsheere, 1981).

9 Links Between Coding Categories and Target Profiles

Even if the coding categories of the grids are very synthetic, we can nevertheless code the animators' and the users' target behaviour.

9.1 Limits of coding

All the target animator's behaviour can be coded using the grid. Nevertheless some different aspects of behaviour are coded in the same category (e.g. elicit the variation of an action and elicit the production of an effect). We do not distinguish between resources according to whether they are personal (e.g. user's notes) or not (e.g. reference guides).

Sometimes, a category includes a target and non-target behaviour. For instance, the category "description of other's action" includes the positive description of the user's action and additional description that are not recommended to be done systematically if we want the user to describe his (her) problem solving process him(her)self.

Behaviour linked to social interaction between users is often confused with individual behaviour (e.g. coding does not determine whether the user describes his (her) own action or other's). But these limits are linked to the need for a simple grid to code the activity directly.

9.2 Coding of non target behaviours

A few behaviours are not found in the animator's target profile:
- elicitations of well defined action,
- explanations (oral, written, ...),
- negative evaluations,
- agreement.

Other non target behaviours are coded at reduced levels on the grid (e.g. elicit a management behaviour is coded "elicit an action").

Some categories are not included in the user's target profile:
- call the animator,
- listen to the animator,
- elicit something,
- answer.

Nevertheless these categories provide information on the learners' cooperative activity, that make it possible to answer questions such as who is the leader? Who asks questions? Who answers?...

As has been done for the animator, other non target behaviours are coded at reduced levels on the grid.

10 Context of The Experiment

The observations took place in a primary school in 1988. The teachers and pupils had some experience with LOGO. Control technology had been introduced in 5th grade.

10.1 Hardware and software

Computers were Apple IIe and Commodore 64. Some routines written in LOGO (french versions available on these computers) made it possible to control the robots (Sougné, 1988). The robotics material consisted of LEGO® sets.

10.2 Users

The pupils were 10 or 11 years old. They had worked with LOGO for four months (about 20 hours with turtle graphics). A few of them (mainly boys) were familiar with LEGO® Technics material the components of which are very close to the robotics sets.

First the pupils arranged themselves in single-sex groups of three or four. After a few periods some preferred to work alone or in pairs. This phenomena seems to be linked with the characteristics of the task: it is not easy to build one robot together because only one pupil can put the pieces together. Task repartition seems easier when there is a computing activity (e.g. one user types at the keyboard, the other takes notes). Nevertheless, some learners continued to work together.

10.2.1 Hypotheses

Control technology activity should be characterized
   * during the building phase of the robots, by a lot of activity with the robotics material, descriptions and explanations of it and
   * during the programming phase by a lot of actions on the computer, descriptions and explanations about them.
- Then, at each phase, the category "Action" should represent a large part (at least 1/3) of the learners' behaviour. These actions should be focussed on robotics material and on the computer (H1).
- Because there are documents related to building robots consulting and to computing activity, we should observe whether learners consult them. The consulting of reference guides should not occur during the building phase if learners have their own projects (H2).
- Girls' profiles should be different from those of boys' (H3).

10.3 Animators

There were two animators. Their profiles of intervention had already been observed in another research project where we tried to regulate their interventions.

10.3.1 Hypotheses

Even where control technology is a new activity for these animators, we can formulate the hypothesis that their general profiles should be similar to those we already have observed (H4). In fact, building and controlling robot should result in animation behaviour we can observe using the grid, such as:
- elicitations of actions,
- elicitation of description of action,
- elicitation of explanation,
- explanations,
- encouragements, reinforcements from the animators.

Another hypothesis is that certain of the animator's behaviour should occur more often when he/she is called by the learners:
- elicit the consultation of reference guide (H5);
- elicit why an action has produced an effect (H6);
- give explanations (H7);
- invite to act (H8).

10.4 Observers

10.4.1 Role

Two researchers were in the school to help in the launching of the project. After a few weeks of activity they decided, with the agreement of the teachers, to code the learners' behaviour and on occasion the teachers' behaviour.

The goal was to provide animators with information to help them regulate their intervention. Therefore it is external-observation with a external-regulation (an observer gives and interprets the information captured using the grid). We also studied whether animators are able to interpret the data to make (if necessary) regulation decisions.

10.4.2 Observers' training

The observers had considerable experience of coding. One of them is the creator of the grid, the second had already coded some LOGO activities. Coding agreement between the observers had been studied. These observers generally noted the same rate of behaviour and coded 90 % of the behaviour in the same categories (an excellent score!).

11 Data collection

11.1 Animators

Animators were observed for the past four periods of activity (80 minutes) of the year. We made 11 observations. Their durations varied for 5 to 12 minutes per animator, depending on the availability of the observers (who are also resources persons for teachers and pupils during the activity). The first animator (M) was observed for a total of 41 minutes and the second one (P) for 67 minutes. The data reflects only some of the animators' behaviour during a control technology activity.

	M	P
N observation periods	5	6
Total duration (minutes)	41'	67'

Fig. 2. Observation of animators

11.2 Users

The thirteen children were observed during the building robot activity. Learners who worked alone were not observed frequently. We here report only observations of two learner trios because we observed them several times. Three girls built a merry-go-round and controled it with the computer. Three boys programmed a crane. The following chart presents the spread of the observations. The unit of observation is one minute.

Names of children	Girls			Boys		
	I	C	N	K	A	G
Observation 1	5	5	5	5	5	5
Observation 2	10	10	10	10	10	10
Observation 3	10	10	-	5	5	5
Observation 4	10	10	10	2	2	2
Observation 5	10	10	10	-	11	11
Observation 6	-	4	4	-	-	-
Total (minutes)	45	49	35	22	33	33

N.B.: - means 'absent'

Fig. 3. Observations of the users

12 Results

12.1 Animators' profiles

12.1.1 Average frequency of interventions

The total number of items of behaviour were 305 for P and 145 for M. P's average was 4,5 / minute and M's was 3,5/minute.

12.1.2 General profile

Most frequent behaviours are elicitations, especially of verbalizations.

A large part of their behaviour consists of describing learners' actions or giving them information.

The animators evaluate and encourage the learners more frequently than they ask them to act on the computing, robotics or didactic material.

They rarely have conversations which do not refer directly to the activity.

	P NP= 305		M Nm= 143	
	N	%	N	%
NON CODING BEHAVIOURS	9	2,95%	8	5,59%
ELICITATIONS	161	53,00%	63	44,00%
ACTIONS ELICITATIONS	40	13,11%	26	18,18%
Elicits an action on the computer	16	5,25%	10	6,99%
Elicits an action on robotics material	9	2,95%	0	0,00%
Elicits an action on reference documents	4	1,31%	8	5,59%
Elicits note keeping	6	1,97%	6	4,20%
Elicits reading notes	5	1,64%	2	1,40%
Elicits social interactions	0	0,00%	0	0,00%
VERBALIZATIONS ELICITATIONS	121	39,67%	37	25,87%
Elicits the description of past action	12	3,93%	3	2,10%
Elicits the description of future action	34	11,15%	12	8,39%
Elicits the dependency relation (why)	20	6,56%	6	4,20%
Elicits a prediction of the action result	19	6,23%	7	4,90%
Elicits the decomposition of action	13	4,26%	4	2,80%
Elicits how to obtain an effect	18	5,90%	3	2,10%
Elicits the evaluation of action/project	5	1,64%	2	1,40%
DISTRIBUTION OF INFORMATION	76	24,92%	43	30,07%
Describes other's action	29	9,51%	8	5,59%
Gives some explanations	40	13,11%	31	21,68%
Agrees	7	2,30%	4	2,80%
EVALUATIONS & ENCOURAGEMENTS	59	19,34%	29	20,28%
Evaluates positively	36	11,80%	9	6,29%
Evaluates negatively	4	1,31%	5	3,50%
Invites to act	19	6,23%	15	10,49%
Total	305		143	

Fig. 4. Animator's profiles

There are differences between P's and M's profiles (Fig. 5):
- P elicits the users more often than M (52,7 % versus 44 %).
- M gives more information than P (30 % versus 24,9 %).

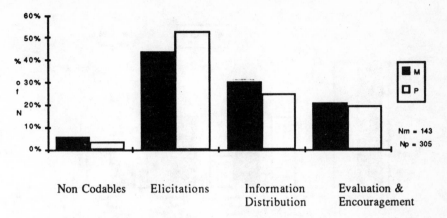

Fig. 5. General animator's profile

12.1.3 Detailed profiles

The study of the spread of animators' behaviours inside the coding categories helps us to analyse inter-animator differences.

Elicitations. About half of M's and P's behaviour belongs to this category (52,7 % of Np and 44 % of Nm.). Verbalization elicitations are more frequent than action elicitations, especially with P for whom this kind of elicitation represents 3/4 of his elicitations (Fig. 6).

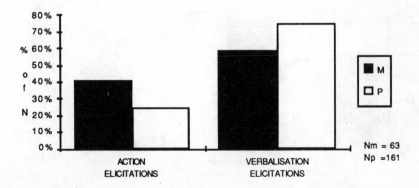

Fig. 6. Comparison between action and verbalisation elicitations

The action elicitations (A.E.) are, in general, proportionally more important for M than P. Nevertheless P elicits action on robotics material (22,5 % of his action elicitations) whereas M never does so (Fig. 7).

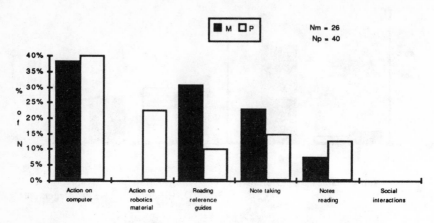

Fig. 7. Animator's behaviours inside "action elicitations"

Certain differences between M and P appear as follows:
- elicitation to consult references guides (30,7 % of AE for M versus 10 % for P),
- asking to make notes (23 % of AE for M versus 15 % for P),
- elicitation to read notes (8 % of AE for M versus 13 % for P).

M and P never elicited cooperation between pairs, neither working together nor focusing on the other's point of view.

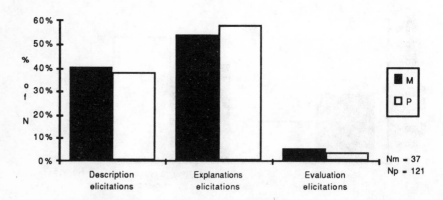

Fig. 8. Animator's verbalisations elicitations

P elicited more verbalizations (39,6 % of Np) than M (26,1 % of Nm). Which is the distribution of those elicitations (V.E.) in the sub-categories of the grid (Fig. 8)?

P elicits more explanations than M. This explains in large part the difference between P and M concerning verbalization elicitations. Explanation elicitations are generally more frequent than description ones. Both animators rarely elicit evaluation of learners' action or project.

For both animators, description elicitations (figure 9) are the most often related to action to be produced (74 % for P and 80 % for M.)

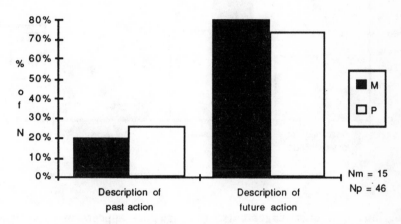

Fig. 9. Repartition of description elicitations

Each type of explanation generally represents 15 to 35 % of the explanation elicitations. It concerns relation between effect and action, decomposition of action, description of action to be produced to obtain an effect and prediction of the result of action. Prediction elicitation is higher for M than P. But P asks more often than P for a description of the action process to produce an effect (Fig. 10).

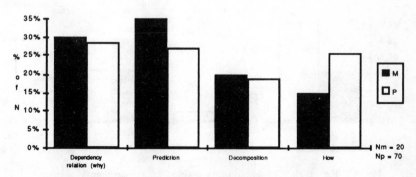

Fig. 10. Repartition of explanation elicitations

Information. The rate of this kind of behaviour is almost the same in P's and M's profiles (24,9 % of Np and 30 % of Nm).

Differences between P and M (Fig. 11) are related to:
- a very high rate of explanations given by M to the learners (72 % versus 52 % for P),
- a higher rate of description of other's action for P (38,1 % versus 18,6 % for M).

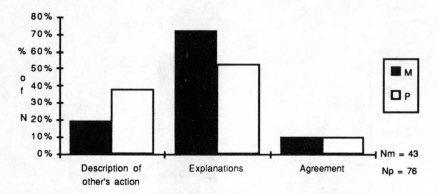

Fig. 11. Repartition of behaviours inside "Information" category

Evaluations and encouragements. This category appears in equivalent proportions in P's and M's general profiles.

M's profile (Fig. 12) is characterized by:
- a larger number of
  * invitations to act (51,7 vs 32,2 %)
  * negative evaluations (17,2 vs 6,7 %),
- fewer positive evaluations (31 vs 61 %).

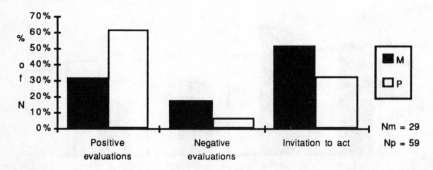

Fig. 12. Repartition of behaviours inside "encouragements" category

12.1.4 Target profile of animation

There is a high percentage of target behaviours. They represent:
- 57,3 % of M's profile,
- 63,9 % of P's profile.

The differences in target behaviour between P and M is due to the fact that P elicits more explanations than M.

12.1.5 Comparison between spontaneous and elicited interventions

If we refer to our previous hypotheses (H5 to H8), some kinds of animator behaviour should be emitted more often when the learners call the animator, these are:
- elicit the consultation of reference guides,
- elicit why an action has produced an effect,
- give explanations,
- invite to act.

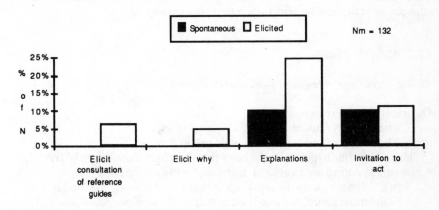

Fig. 13. Spontaneous and elicited behaviours (animator M)

These hypotheses (H5 to H8) appear to be confirmed only for M. Most often this animator elicits the consulting of didactic documents and asks for explanations of the relation between an action and an effect only when learners have called him. He then gives them an explanation and invites them to go on.

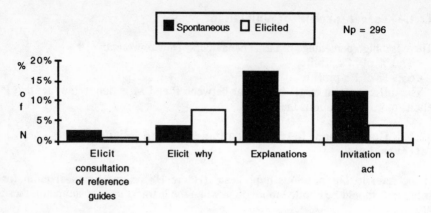

Fig. 14. Spontaneous and elicited behaviours (animator P)

P, however, elicits the consultating of books, gives explanations and invites the learners to act more often when his intervention is spontaneous. Only hypothesis H6 is confirmed for P: most frequently he elicits the description of a relation between an action and an effect when the learners call him.

12.2 Users' profiles

12.2.1 Average frequency of interactions

On average, learners have about
- 3 behaviours / minute in the male group *(mg)*,
- 6 behaviours / minute in the female group *(fg)*.

This follows our hypothesis: the boys' profile differs from the girls' (H3).
Nevertheless there are individual differences within the groups:
- In the female group, I's rate is on average 1,5 / minute; S 3; and N 1,4.
- In the male group, K's rate is on average 1,4 / minute; A 0,7; and G 1,1.

So there are at least as much individual variation as gender variation (- H3).

12.2.2 General learners' profiles

Learners' interactions are task centered. Only about 6 % of their behaviour is not (5,8 %, in the female group, 6,1 % in the male group). About a third of learner behaviour is made up of actions (36,5 % among the girls and 32,6 % among the boys). Users call and listen to the animators (7,2 % of N *fg* and 9,1 % of N *mg*).

Half of learners' behaviour consists of verbalizations (50,3 % of N *fg* and 53 % of N *mg*).

At this level, there is no gender difference within the global distribution of behaviour types.

	I		S		N		K		A		G	
	N	%	N	%	N	%	N	%	N	%	N	%
Actions	**22**	**32%**	**56**	**40%**	**18**	**33%**	**9**	**28%**	**7**	**32%**	**16**	**41%**
Action on the robotics material	10	15%	28	20%	13	24%	7	22%	6	27%	10	26%
action on the computer	3	4%	23	16%	4	7%	1	3%	1	5%	6	15%
Action on documents	6	9%	3	2%	0	0%	0	0%	0	0%	0	0%
Action on notes	3	4%	2	1%	1	2%	1	3%	0	0%	0	0%
Elicit the animator	**3**	**4%**	**3**	**2%**	**2**	**4%**	**2**	**6%**	**0**	**0%**	**1**	**3%**
Listen to the animator	**4**	**6%**	**6**	**4%**	**3**	**5%**	**3**	**9%**	**2**	**9%**	**1**	**3%**
Elicitations	**14**	**21%**	**24**	**17%**	**16**	**29%**	**7**	**22%**	**5**	**23%**	**3**	**8%**
Elicit evaluation	1	1%	2	1%	3	5%	1	3%	0	0%	0	0%
Elicit agreement	0	0%	3	2%	3	5%	2	6%	0	0%	2	5%
Elicit action on the computer	8	12%	10	7%	1	2%	1	3%	2	9%	0	0%
Elicit action on documents	0	0%	0	0%	0	0%	0	0%	1	5%	0	0%
Elicit action on robotics material	0	0%	2	1%	4	7%	2	6%	0	0%	1	3%
Elicit how to do something	2	3%	2	1%	1	2%	0	0%	1	5%	0	0%
Elicit why	0	0%	0	0%	0	0%	0	0%	0	0%	0	0%
Elicit an action description	1	1%	1	1%	2	4%	0	0%	1	5%	0	0%
Elicit the description of project	2	3%	4	3%	2	4%	1	3%	0	0%	0	0%
Descriptions	**24**	**35%**	**46**	**33%**	**15**	**27%**	**8**	**25%**	**8**	**36%**	**17**	**44%**
Describe future action (project)	12	18%	20	14%	9	16%	7	22%	1	5%	9	23%
Describe the past action	2	3%	7	5%	0	0%	0	0%	1	5%	1	3%
Describe how to do something	0	0%	3	2%	0	0%	0	0%	4	18%	2	5%
Describe why	0	0%	1	1%	0	0%	0	0%	0	0%	1	3%
Evaluate action or idea	9	13%	12	9%	6	11%	1	3%	2	9%	4	10%
Predict the action result	1	1%	3	2%	0	0%	0	0%	0	0%	0	0%
Answers	**1**	**1%**	**5**	**4%**	**1**	**2%**	**3**	**9%**	**0**	**0%**	**1**	**3%**
Yes	1	1%	3	2%	0	0%	0	0%	0	0%	0	0%
No	0	0%	2	1%	1	2%	3	9%	0	0%	1	3%
TOTAL	**68**		**140**		**55**		**32**		**22**		**39**	

Fig. 15. General learners' profiles

12.2.3 Detailed learners' profile

We have not distinguished the originators of off task behaviour. So we shall not refer to them hereafter. This makes the group proportions as follows:

Female group	Male group
Ni = 68	Nk = 32
Ns = 150	Na = 22
Nn = 55	Ng = 38

Female group (I, S, N). On average, S is twice as active than her partners. She types frequently on the keyboard (15,3 % of Ns versus 4,4 % of Ni and 7,2 % of Nn). We should note that only one user can do this at a given moment; consequently the others are less active (at least to our observation).

Only S describes the strategies necessary (how) to obtaining a desired result (2% of Ns). I, S and N rarely predict the results of their action (1,4 % of Ni, 2 % of Ns and 0 % of Nn). Nevertheless the three users frequently describe their projects or the instruction to be typed at the keyboard (17,6 % of Ni, 16 % of Ns and 16,3 % of Nn). Only I and S describe the action that has been produced (1,9 % of Ni and 4,6 % of Ns). S only once verbalises the relation between an action and the effect obtained.

The three girls frequently evaluate their actions and ideas (13,2 % of Ni, 8 % of Ns and 0,9 % of Nm). The proportion of actions on robotics materials is important. It represents 14,7 % of Ni, 18,7 % of Ns and 23,6 % of Nn.

Users sometimes take notes (4,4 % of Ni, 1,3 % of Ns and 1,8 % of Nn) and they consult reference books, except N (4,4 % of Ni and 2 % of Ns) (+H2). The largest part of their actions are then actions on the computer and on robotics material (+H1). Learners call the animators (4,4 % of Ni, 2 % of Ns and 3,6 % of Nn) and listen to their explanations (5,8 % of Ni, 4 % of Ns and 5,4 % of Nn). Interactions between animators and users were a little bit more frequent when child I was absent. This might confirm her leadership role.

Most of the elicitations deal with an action with the computer (11,7 % of Ni, 6,6 % of Ns and 1,8 % of Nn). We can imagine that S' elicitations didn't have the expected effect on her partners because their frequency is higher (N = 10) than her partners' actions on the computer (N = 7). S and N sometimes elicit actions on the robotics material.

Learners never elicit action on the didactic materials nor the description of the link between an action and an effect (why). But they elicit descriptions of their projects from each other, how to realize them and past action. They ask information on primitives, elicit evaluation and the agreement of their partners.

Male group (K, A and G). About a quarter of the activity of this group consists of building their robot (21,8 % of Nk, 27,2 % of Na and 26,3 % of Ng). Learners sometimes type on the keyboard, especially G (3 % of Nk, 5 % of Na and 15 % of Ng). They do not consult guides and do not take notes, except M who does it once. It is not really surprising because they are focussed on their construction, a vehicle that they build independently, without model (+H2). Another hypothesis is also confirmed with these learners: the 'action' category represents a large part of their learner' profile and actions are focussed on robotics materials and the computer. These children call the animator and listen to him (+H1).

K and A make more elicitations than G (22 % of Nk, 23 % of Na versus 8 % of Ng). Elicitations vary in their nature, they also vary between users. These three never ask about the dependency relation between their action and the computer's reaction.

Users (except A) often describe their intended action (16 % of Nk, 5 % of Na and 23 % of Ng). Nevertheless A describes more frequently than his partners, how to produce an effect (18 % of Na versus 0 % of Nk and 5 % of Ng).

Learners evaluate their actions and ideas (3 % of Nk, 5 % of Na and 10 % of Ng) but never predict their results.

They rarely describe an action they have just carried out and its relation to the observable effect produced.

We never observed expressions of approval from the three boys. K and G sometimes expressed disapproval of an idea or an action of their partners (9 % of Nk and 3 % of Ng).

12.2.4 Learners' target profile

How much is the users' target profile reflected in his/her actual profile ?

This varies from one user to another. The target behaviour represents 28,1 % to 48,5 % of the learner profiles.

The differences observed are greater between individuals than across gender (- H3). For the female group, target behaviours were:
- 48,5 % of Ni;
- 34 % of Ns
- 29,1 % of Nn.

In the male group, we notice the same variations. The target behaviours were:
- 28,1 % of Nk
- 36,3 % of Na
- 44,7 % of Ng

12.2.5 Regulation decisions

What decisions might we take if these kinds of observations were repeated? Our goal should be to lead the animators to elicit target behaviour which seldom or never appeared in learners' target profile.

Female group (I, S, N). During the six observation periods, we rarely (or never) observed the following behaviour:
- predictions (the three girls),
- relation between action and effect produced (for I),
- explanation of how to produce an effect (for I and N and - but less - for S),
- descriptions of past action (for N),
- consulting of guides (for N and S),
- reading or taking notes (for N and S).

Male group (M, A and G). During the five observation periods, the following behaviours rarely (or never) occured:
- consulting reference guides,
- writing or reading notes,
- description of past action,
- description of the dependence relation between an action and the effect produced,
- prediction,
- explanation of how to produce an effect (in the case of M).

12.3 Effects of the communication of the results of the observations

Part of the work of the observers was to provide a focus for discussion of the teaching and learning processes.

The discussion was focussed on animators' behaviour because we had not, at this time, analysed the learners' data.

12.3.1 Teaching behaviours

After a few observations, the results were communicated to the animators. These one of the observers helped them interpret the data.

General feelings. In general, animators were satisfied about their profiles, with the exception of socialization: there was a lack of cooperation between learners. This phenomena was also felt independently of the results of observations. Only in one group (the female one) did the members develop their initial project together. A second trio (the male group) was beginning to work together when we began the observations. Other pupils often worked alone.

Elicit cooperation. Even where perceived a cooperation problem without the help of the results of the observations, the animators did not elicit social interactions such as cooperation or confrontation of points of views. However, this type of intervention is one of their objectives and the animators have recently reported their intention to reinforce cooperation between learners. (There is no operational decision coming from general feelings.)

Following the observations, strategies to increase cooperation between learners were discussed. The outcome was a proposal to combine all the models already built (crane, convey belt, vehicle). For instance, the crane would put pieces on the convey belt which would convey them to a vehicle which would then take them to their destination. If the learners accepted this project, cooperation would clearly be increased.

Give less information. Taking note of the observations, M decided to try to restrict his giving explanations only to times when learners can not find the information themselves.

Congruence and realism related to the animators' profiles. Ignoring the problem of cooperation and the fact that M gives sometimes a lot of information to the learners, we observed that a large part of the animators' behaviour belongs to the target profile. That is, there is a good congruence between the target and the observed behaviour.

Moreover, animators seem to be conscious of their kinds of intervention. They know they very often elicit responses from the learners and that their elicitations are focussed on explanations of users' mental processes.

Nevertheless, they say that, independently of their target profile, their role is also to help the learners to solve their problems, even if they help them a lot.

12.3.2 Users' behaviours

Congruence between target and observed users' behaviours. The general feelings of the animators about the observed pupils are positive: learners of each group being involved in their own project and problem solving.

Nevertheless, the results of the observations have shown that some target behaviour is rarely or never produced. However, because these results only sample part of the activity, it is impossible to know if theses observations would have been repeated during a longer observation period.

The observation helps to draw attention to some facts even if they can not be generalized.

The communication of the observation results should have implied regulation decisions, but we did not have the opportunity to apply them because it was the end of the school year.

13 Conclusions

Coding of an animator's and of users' behaviour with direct observation grids is possible.

With regular, but not necessary long, observation we are able to establish general animator and user profiles. The interpretation of the results helped the animators to be more conscious of their interventions and to evaluate whether the users had reached certain specific objectives (here focussed especially on problem solving and socialization). Thus this information helped the animators to regulate their intervention. Nevertheless, we have to remember that the results cannot be generalized. If we want to be sure that the data are representative of the observed activity before deciding to regulate it, we have to try to master a maximum of variables (e.g. moment of observation, duration, reliability of coding, ...). We can use experimental designs such as multiple baseline schedules where there are at least three observations with a certain constancy in the target population's behaviour.

A limitation of direct coding is that we cannot contextualize the interactions and then make sequential analysis. It is, then, impossible to study directly the efficiency of animator's intervention. Moreover, with direct coding, we loose information because it is very difficult to catch all the behaviour and to code it directly. So this observation method gives some qualitative indicators (observed or not) but not real quantitative results. But what is important is to get indicators that can help the animators to think about their own and the users' profiles and to regulate them.

Our first hypothesis was that an animator's training in using observation grids could help them to be more congruent and realistic and to self-regulate their intervention. Here we have seen that assisted regulation has occured with the help of external observers. Observations often confirmed the animator's feelings. The most important contribution of observators to regulation was to provide a special time for checking and decision making.

Other researches have demonstrated that self-regulation is possible (Denis, 1990; Gilbert, 1990); even if it is very difficult for a person to analyse an educational situation in which he/she is involved. Providing an information to the animators helps them to do this. Nevertheless this analysis is limited because it is only based on some explicit animator's and users' behaviours. It does not take into account many of the factors which intervene in a learning situation (e.g. animator's representations, contents, ...) or certain effects of animator's intervention (especially long term effects). Nevertheless, we can hope that our tools would help animators to individualize their interventions more and to become more professional.

Acknowledgements

Thanks to Professor D. Leclercq for his critical point of view on this research and for the re-reading of the first draft of this paper. Thanks also to Mike P. Doyle for his complete re-reading of this text.

References

Battro, A.M., *Dictionnaire d'épistémologie génétique*, Dordrecht, Holland, 1966.

Bayer, E., *Une science de l'enseignement est-elle possible ?* In: M. Crahay, D. Lafontaine (1986), pp. 483-507.

Black, J.B., Swan, K., Schwartz, D.L., Developing thinking skills with computers, *Teachers College Record*, Columbia University, 89 (3), 1988.

Bloom, B.S., *Caractéristiques individuelles et apprentissages scolaires*, Bruxelles, Labor, 1979.

Cardinet, *Evaluer les conditions d'apprentissage des élèves plutôt que leurs résultats*, communication colloque rencontre belgo-suisse, Namur, 1983.

Charlier, E., Donnay, J., Un enseignant, un décideur. In: *Formation Recherche en Education*, pp. 3-10, 1985.

Clark, C.M., Peterson, P.L., *Teachers' thought processes*, in M.C. Wittrock (ed.), *Third Handbook of Research on Teaching*, New York, Macmillan, pp. 255-296, 1987.

Clements, D.H., Nastasi, B.K., Social and cognitive interactions in educational computer environments, *American Educational Research Journal*, 25 (1), 1998, pp. 87-106.

Clements, D.H., Effects of LOGO and CAI Environments on Cognition and Creativity, *Journal of Educational Psychology*, 78 (4), 309-318, 1986.

Clements, D.H., Gullo, D.F., Effects of Computer Programming on Young Children's Cognition, *Journal of Educational Psychology*, 76, 1051-1058, 1984.

Clements, D.H., Longitudinal study of the effects of Logo programming on cognitive abilities and achievement, *Journal of Educational Computing Research*, 3 (1), 1987.

Close et Butler, *Children's problem solving processes in LOGO*, communication at congress EUROLOG, Dublin, 3-6 September 1987.

Crahay, M., *De l'épistémologie génétique à l'action éducative*, Paris, PUF, 1989.

Crahay, M., Lafontaine, D. (eds), *L'art et la science de l'enseignement*, Bruxelles, Editions Labor, 1986 (coll. Education 2000).

De Landsheere, G., E. Bayer, *Comment les maîtres enseignent. Analyse des interactions verbales*, Bruxelles, Direction générale de l'Organisation des Etudes, Pédagogie et Recherche, n°1, 1981, 4e éd.

De Landsheere, G., *Introduction à la recherche en éducation*, Liège, G. Thone, 1976, 4e édition.

Denis, B., *Manuel LOGOWRITER illustré.* Liège, Service de Technologie de l'Education de l'Université, 1988.

Denis, B., *Quels sont les projets privilégiés par les utilisateurs de LOGO?* Liège, Service de Technologie de l'Education de l'Université, 1988.

Denis, B., Robotique pédagogique et formation des enseignants, communication à l'Université d'Eté de la Commission des Communautés Européennes, Ghent, September 1988.

Denis, B., Technologie de contrôle et LOGO. La robotique, ses enjeux, ses modalités, in *Education-Tribune Libre,* September 1987, 208, pp. 61-67.

Denis, B., *Vers une auto-régulation des conduites d'animation en milieu LOGO, Université de Liège,* Thèse de doctorat non publiée, 1990.

Denis, B., Hardy, J.-L., *Programming action analysis in LOGO for environmental achievement evaluation,* Liège, Laboratoire de Pédagogie expérimentale de l'Université, 1985.

Doise, W., Mugny, G., *Le développement social de l'intelligence,* Paris, Inter Editions, 1981.

Doyle, W., Powder, G.A., Classroom ecology: Some concerns about a neglected dimension of research on teaching, *Contemporary Education,* Spring 1975, 156 (3), 183-188.

Doyle, W., Paradigms for research on teaching effectiveness, in L.S. Shulman (ed.), *Review of Research in Education,* (vol. 5), Itasca, III, Peacock, 1978, p. 188.

Doyle, W., *Student mediating responses in teaching effectiveness,* final report, Texas, North Texas State University, Department of Education, 1980.

Emihovich H C., Miller, G.E., Learning Logo: The social context of cognition, *J. Curriculum Studies,* 20 (1), 57-70, 1988.

Emihovich H C., Miller, G.E., Effects of Logo and CAI on black first grader's achievement reflectivity, and self-esteem, *The Elementary School Journal,* 88 (5), 1988.

Eyre, R., Control technology in the classroom, *Journal of British LOGO User's group,* Spring 1989, pp. 5-7.

Fischer, K.W., A theory of cognitive development: the control and construction of hierarchies of skills, *Psychological Review,* 1980, 87, 477-531.

Flanders, N.A., *Interaction analysis in the classroom: a manual for observers,* Ann Arbor, University of Michigan, 1966.

Gage, N.L., *Comment tirer un meilleur parti des recherches sur les processus d'enseignement?* In: M. Crahay, D. Lafontaine (1986), pp. 411-434.

Gilbert, A, *Formation d'animateurs à l'utilisation de grilles d'observation en vue de réguler leurs conduites,* Université de Liège, Mémoire de licence en sciences de l'éducation, 1990.

Giordan, A., *Une pédagogie pour les sciences expérimentales,* Paris, Centurion, 1978, Collection Paidoguides.

Giovannini, M.L., Denis, B., Sougné, J., *Projet pilote LEGO-LOGO,* Rapport, Universita degli Studi di Bologna (Italia), Université de Liège, 1988.

Giovannini, M.-L., Lodini, E., *Projet I.D.A.: expérience de Bologne,* Communication à l'Université d'été des Communautés européennes, "les N.T.I. et l'enseignement primaire", Liège, 1985.

Hardy, J.-L., *Pourquoi LOGO dans un contexte éducatif?* Bruxelles, Labor, De Boeck, 1985.

Hersen, M., Barlow, D.H., *Single case experimental designs, strategies for studying behavioral change,* New York, Pergamon Press, 1976.

Hoyles, C., Noss, R., *Children working in a structured LOGO environment: from doing to understanding,* Recherches en didactique des Mathématiques, 8, 131-174, 1987.

Leclercq, D., La fonction régulatrice de l'évaluation vue sous l'angle de l'implication de l'étudiant. In: *Education-Tribune Libre*, Bruxelles, n°159, 1976, 65-75.

Leclercq, D., *Quelle technologie de l'éducation dans une société technologique ?*, Document présenté au Congrès International d'Education et de technologie, Vancouver, May 1986.

Leclercq, D., L'ordinateur et les défis de l'apprentissage, *in Horizon*, 13, November 1987.

Leclercq, D., Introduction à la Technologie de l'Education, Liège : Service de Technologie de l'Education, Université de Liège, 1992a.

Leclercq, D., The validity, the reliability, and the sensitivity of self-assessment. In: D. Leclercq, J. Bruno (eds.). Item banking: interactive testing and self-assessment. NATO ASI Series F, Vol. 112. Berlin, Springer-Verlag, 1993.

Matos, F. J., The construction of the Concept of Variable in a LOGO Environment : A Case Study, Proceedings of the Tenth International Conference Psychology of Mathematics Education, London, 1986.

Medley, D.M., *The effectiveness of teachers*. In: Peterson, Walberg, Research on teaching, McCutchan Publishing Corporation, 1979.

Nonnon, P., *Laboratoire d'initiation aux sciences assisté par ordinateur*, Université de Montréal: Faculté des Sciences de l'Education, 1986.

Ocko, S., Resnick, M., Integrating LEGO with LOGO. Making connections with computers and children, *The Media Laboratory*, MIT, Cambridge, MA, 1987.

Orban, M., *La fonction d'enseignant : mutations et perspectives*, Extrait du rapport FAST "Les nouvelles technologiques à l'école". Quelles utilisations de l'ordinateur, Recherche commanditée par le Ministère belge de la Politique scientifique, September 1986.

Papert, S., *Jaillissement de l'esprit*, Paris, Flammarion, 1981.

Papert, S., *Microworlds : transforming education*. Paper presented at the ITT Key Issues Conference held at the Annenberg School of Communications of the University of Southern California, 14 March, 1984.

Papert, S., *Teaching children thinking*, Cambridge, MIT. A.I. Laboratory, LOGO memo n° 2, October 1971.

Piaget, J., *L'équilibration des structures cognitives*, Paris, PUF, EEG, XXXIII, 1975.

Piaget, J., *La psychologie de l'intelligence*, Paris, Colin, 1956.

Resnick, M., *LEGO-LOGO : Learning through and about Design*, The Media Laboratory, MIT, Paper presented at the 1989 AEREA Annual Meeting.

Resnick, M., Ocko S., Papert, S., *LEGO, LOGO and Design*, The Media Laboratory, MIT, Paper presented at the 1989 AEREA Annual Meeting.

Robinson, M.A., Feldman, P., Uhlig, G.E., The effects of Logo in the elementary classroom : an analysis of selected recent dissertation research, University of South Alabama, Mobile, Alabama 36688, in *Education*, 107 (4), 434-442, 1987.

Romainville, M., Une analyse critique de l'initiation à l'informatique : quels apprentissages et quels transferts ? In: *Colloque francophone sur la didactique de l'informatique : Actes - version distribuée aux participants*, Paris, 1-3 September 1988.

Rosenshine, B., Furst, N., Use of direct observation to study teaching. In R. Travers (ed.), *Second Handbook of Research on Teaching*, Rand McNally, 1973.

Séron, X., Lambert, J.-L., Van der Linden, M., *La modification du comportement, Théorie, Pratique, Ethique*, Bruxelles, Mardaga, 1977.

Sougné, J., *Les primitives LEGO-LOGO français (logiciel)*, Liège, Service de Technologie de l'Education de l'Université, 1989.

Sougné, J., LOGO-SCAN, A tool kit to analyse LOGO programs. Paper presented at ICTE, March 1990, 313-315.

Tetenbaum, T.J., Mulkeen, T.A., LOGO and the teaching of problem solving : A call for a moratorium, *Educational Technology*, November 1984.

Texier, A., *Des ailes pour la tortue,* Paris, Eyrolles, 1986.

Thirion, A.-M., Evaluation de la recherche-action. In: *Problèmes de l'éducation préscolaire,* Strasbourg, Conseil de l'Europe, 1975, pp.232-243.

Valcke, M., *The integration of Logo programming in the curriculum of the primary school, outline and example of an innovation strategy,* EDIF, State University Ghent.

Walker, D. F., Logo needs research: a response to Papert's paper, *Educational Researcher,* June-July 1987.

Annexes

Annex 1: Animator's observation grid
Annex 2: User's observation grid

UNIVERSITE DE LIEGE
Faculté de Psychologie
et des Sciences de
l'Education
SERVICE DE TECHNOLOGIE
DE L'EDUCATION
D. LECLERCQ

ANIMATORS

CENTRE
de TECHNOLOGIE
de l'EDUCATION

						Interventions	
Non coding behaviors						Spontaneous	Elicited
E L I C I T S		A C T I O N		On computer	well defined		
					effect		
				On robotics material			
				On reference guide or document			
				Note keeping			
		Social interaction, cooperation					
		V E R B A L I Z A T I O N		Description of Action	Evaluation		
					Past action		
					Future action		
					Dependance relation (why)		
				An Explanation	Prediction		
					Decomposition		
					How		
Gives Information		Other's action description					
		An oral explanation or written, on keyboard or robotics material					
		Agrees					
Evaluates Encourages		Positively					
		Negatively					
		Invites to act					

Animator's observation... User's number ....
Observation by.. Number of computer.................... Duration of observation....................

Name of the couples												
Spontaneus Interventions												
Elicited Interventions												

Annex 1. Animator's observation grid

UNIVERSITE DE LIEGE
Faculté de Psychologie
et des Sciences de
l'Education
SERVICE DE TECHNOLOGIE
DE L'EDUCATION
D. LECLERCQ

USERS

CENTRE
de TECHNOLOGIE
de l'EDUCATION

CTE

			X	Y	Name	Name
	General total					
Non coding behaviors	Total 1				X	Y
	Total 0					

A C T I O N S	X	Y	On robotics material			
			On computer, keyboard			
			On reference guide or document			
			On notes	reading and writing		
				screen		
			Calls the animator			
			Listens to the animator			

V E R B A L I Z A T I O N S	X	Y			an opinion, an agreement				
			E L I C I T S	X	Y	An Action	X	Y	well defined on computer
									effect to produce
									on robotics material
									document
						An Explanation			How
									Why
									Prediction
						A Description			Future action
									Past action
	X	Y	D E S C R I B E S	X	Y	An Action	X	Y	Future action
									Past action
						An Explanation			How
									Why
									Prediction
									Decomposition
			Evaluates	Positively					
				Negatively					
			Answers	Yes / No					
				Doesn't Know					

Observation by .... Duration of observation............. from.....................to.....................Date..........
Animator.. Classroom..............................School.................................n° Obs..................
Objectives...
Remarks...

Annex 2. User's observation grid

3. Tools Developed for Control Technology

3.1 State of the Art

Workshops

Reporter: Mike P. Doyle, Honorary Chairman, **Logo**S: the Logo User Group

Abstract. A number of practical demonstrations and workshops took place during the NATO ARW on Control Technology in Elementary Education. These workshops are outlined and their content, particularly software, is analysed from the viewpoint of how it represents the reality controlled. A particular device is used to focus the analysis: the language cone.

Keywords. Demonstrations, Floor Robots, Hardware, Interfaces, Language Cone, LEGO®, LOGO, Motors, Overlay Keyboards, Parallel Processes, Software, User Environments, Representation, Workshops.

1 Introduction

The workshop presentations divide themselves into four groups as follows:

Overlay keyboard	Mike Doyle, Duncan Louttit
Floor robots	Duncan Louttit, Maurice Meredith
Modelling and control	Eduardo Calabrese, Jean-Baptiste La Palme, David Argles, Ole Møller.
Control and context	Martial Vivet and Pascal Leroux

The last mentioned workshop consisted of more than a particular item of hardware or software. It was an opportunity to participate in a control technology based pedagogical environment. This has been summarised by its presenters. Their account follows the report on the other workshops.

1.1 A viewing framework

It is possible to take a variety of viewpoints on these workshops. We here consider them from the unifying viewpoint of "representation". In all cases the object controlled is a simple assemblage of motors, switches, lights, etc. Each workshop provided a different viewpoint on the control, via a computer, of a model constructed of such components. The model itself might have been a model train, a LEGO® or Fischer-Technik® construction, a ready-made robot, or a scratch built model.

It will help clarify our viewpoint if we put on a version of Pierre Nonnon's cognitive spectacles. Let me explain a little, Fig. 1 will help.

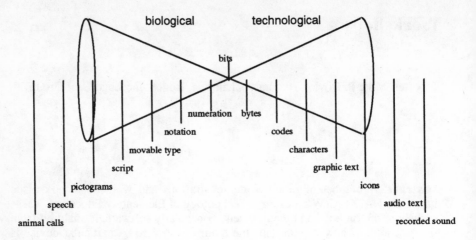

Figure 1. The language cone

If we include within the left cone the world of the concrete model and the people who manipulate it, and include within the right cone the world of computer representation, then we have a framework for discussing all the workshops.

The right cone represents the development of written language up to the point where it has been reduced to a sequence of binary digits which may be transformed as they pass through electronic logic. But, before we proceed to discuss the workshops themselves we must understand a little about the left cone.

The left cone represents a representation of the reality on the right in written form. (By written form we here mean not the writing in natural language with which the novelist constructs alternative realities within our consciousness, but written instructions in computer language which produce concrete, sensible representations.) We may clarify this notion by reference to Pierre Nonnon's presentation (this volume).

1.2 Levels of representation

Nonnon demonstrated the contiguous occurrence of an ongoing action with a graphical representation of that action. This was achieved by: collecting data from the real-world event; converting that data into binary representation; processing that data according to mathematical rules expressed in computer language; then displaying that data on a representation of a graph on a computer screen, the graph itself being the product of a description of a graph in computer language. The success of the graphical representation in focusing the observer on the critical features of the real-word event is a measure of the effectiveness, and appropriateness, of the representation. In other words, has the computerist succeeded in writing a good description of the world?

The eye of the byte. In discussing the workshops we need to keep in mind that there is continuity of representation "through the eye of the byte". From reality to the simulation of reality. In an educational context, we must ensure that this continuity is not hidden from the learners at the same time as protecting them from intellectual (cognitive) overload whilst they are learning. It was the aim of all the workshop presenters to help bring these two images, through different lenses, together into a single, coherent cognitive experience. The question we need to address is, "How well might they have achieved this for elementary school children?".

2 The Overlay Keyboard

Both workshop presenters demonstrated second generation products in this area.

2.1 Historical development

The educational overlay keyboard originated in the special needs field. In the UK the first design was a membrane "expanded typewriter" design with a double (Apple II) keyboard layout. It was developed to help a physically handicapped pupil program. It was quickly realised that, for a computer using ASCII coding, a simple 8x16 matrix would be simpler to manufacture and potentially more versatile. The original A4, 128 key "Concept Keyboard" was of this form; each key being 18mm square.

2.1.1 Applications

Within a short period of its introduction it became clear that the overlay keyboard was a very versatile input device. Not only was is a means of inputting ASCII encoded text and keyboard control characters but, given suitable software, it could be used to send message - whole strings of characters - to the computer. For younger children it was used to input whole words at a single touch, more appropriate than typing letter by letter; Doyle (1986) demonstrated its power (in conjunction with text-to-speech synthesis) to make educational simulation software more efficient; and a framework program, *Touch Explorer*, provided a very versatile interface to investigative database work. More recently, LCSI (les systèmes d'ordinateur LOGO inc., Quebec) developed LOGO primitives to talk to the overlay keyboard (Doyle, 1991).

2.1.2 Product development

A couple of years after the development of the original A4 unit, an A3 unit was produced. Again, 128 keys were provided. Their layout, however, was different. The, inelegant A4 layout proved unacceptable when enlarged to A3. The new layout had rectangular keys 24mm x 30mm. This both provided a greater active area and constrained the keyboard to landscape use.

2.1.3 Low power units

At the beginning of the 1990's a significant change in design took place. Low-power electronics were introduced. This made possible the powering of the Concept keyboard from a conventional serial port, cf. a mouse. It also enabled the development of intelligent keyboards, the characteristics of which could be changed through software. Current keyboards can return the x,y coordinates of the key pressed and hence act as positional devices, like a mouse or touch screen. The layout designed for the A3 keyboard was now adopted for the A4 unit, with each key was divided horizontally to give 128 or 256 keys.

2.1.4 File standards

In the UK, under the aegis of the NCET, a standard file and driver format for overlay keyboards has been developed. This has been implemented for Acorn RISC-OS and MS-DOS machines, and with some variation for the Macintosh.

2.1.5 New keyboards

The two overlay keyboards demonstrated represent the next stage in development. The design of all the earlier keyboards had been focused on the "bit". The numbers of keys provided had been based on the number of bits available in 7 and 8 bit bytes, (128 and 256 keys). If we look at the language cone we see that this represents thinking close to the binary. Only as an afterthought was the active key area made aesthetically pleasing, though at the cost of reduced versatility.

2.2 The 192 key layout

The 192 key keyboard was developed by working from a higher level, that corresponding to graphic text/icons. That is, the ergonomics were considered first and the electronics derived from this. The design constraints were straightforward:
the layout should efficiently cover two paper sizes, European A4 and US Letter;
each key should be of "standard finger" size;
the keys should be square.
The 12x16 matrix of 16mm square keys used by the Concept 192 (Fig.2) conforms to these requirements (16mm square is the size of a LEGO® brick). It also exactly covers the screen of a standard, 3 x 4 aspect ratio, 14" computer monitor. This means that overlays designed for this overlay keyboard may alternatively be used on a detachable touch screen, such as the Touch Window.

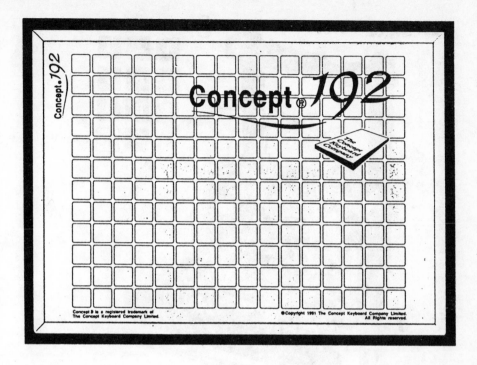

Figure 2. The Concept192 overlay keyboard

The demonstrations which followed exhibited the beneficial effects of shifting the focus of our perception from the engineer's bits to the user's graphics.

2.2.1 The Concept Universal 2/192

Mike Doyle demonstrated a variety uses for the Concept 192 keyboard in conjunction with the LCSI/LEGO LogoWriter Robotics software. Applications variously used the keyboard in portrait and landscape orientation. Some have been described elsewhere (Doyle 1991, 1992a, 1992b). An application relevant to the topic of the seminar was its use with LEGO® Technic. The overlay shown, Fig. 3, is an overlay made by enlarging the final page of LEGO project booklet No 8 for Set 9700, with additional legends. It is used both to emulate, for the original LEGO Interface box, the "setup" screen of the new LEGO® Control Lab, and to control the model illustrated. Inset is a setup overlay made by photocopying the top of the LEGO® Interface box itself.

222

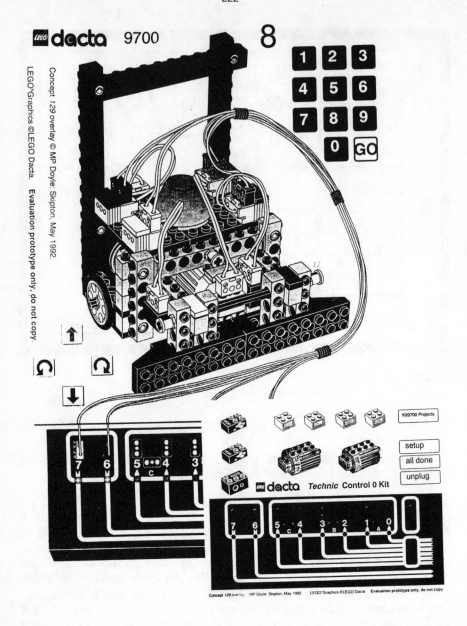

Figure 3. Concept192 overlays for Control LEGO®

2.2.2 Swallow 192

Duncan Louttit demonstrated a 192 overlay keyboard of his own design. Unlike the original Concept keyboard the software for which must reside in the host computer, the Swallow Systems board retains the messages associated with the keys in its own (non-volatile) memory. The keyboard has two modes: "talk" and "teach". When (hardware) switched to teach mode it will associate with any keys pressed the subsequent string typed at the computer keyboard. Switched back to talk mode, the pressing of a programmed key will insert the associated message into the computer's keyboard buffer. Software is required in the host computer to enable this process, but there is no need for look-up tables to interpret the individual character returned by the original Concept keyboard.

3 Floor Robots

Both floor robots on show were inspired by the late Milton Bradley "Bigtrak". This was a programmable futuristic tank-like toy, which was programmed using a small membrane keypad set into its top. It was particularly liked by computer oriented mathematicians, because the manner of its programming and the way it turned and moved it made possible the teaching of the basics of turtle geometry at very low cost. Floor robots are now referred to specifically in the UK national curriculum in the context of computer control.

3.1 Swallow Systems PIP

Named after the Louttit's pet dog, PIP is a minimalist big black brick that can be programmed to move and turn; pause; play tunes; and flash a light. It is programmed via a 4 x 6 keypad mounted on the top rear. PIP's actions are indicated by symbols: arrows for movement; bent arrows for turn; a clock face for pause; a quaver for tone; and a circle with rays for flash. A calculator layout pad is used to enter numerical variables and musical notes. There are RPT and END keys for making repeat loops. Two keys clear memory and the last entry. Finally there is a "GO" button to run the sequence of instructions in memory and a "TEST" button to run an inbuilt program.

3.1.1 Programming PIP

PIP's programming is based in FORTH rather than LOGO. Only one program is possible, without sub-procedures but with repeat loops which may be nested. A program is written by pressing an action key followed by a variable (number/note). A typical program might be: ↑②⑤ ↦④⑤↑②⑤○①⑤↓②⑤↩④⑤↓②⑤ GO, which makes PIP draw a 25cm corner, wait, then retrace its steps. Erroneous entries, e.g. ↑↦, cause PIP to "grumble". There is no other feedback to the user. (Though PIP will also accept a program downloaded into its memory from LogoWriter.)

3.1.2 Control PIP

A prototype control extension to PIP was demonstrated. This PIP had an additional control box containing two state sensors (3, giving a total of 8 states). The sensors were programmed simply by activating them by hand whilst pressing either a movement or the pause button. Thus, to program PIP to back away from an object it had touched, the sensor might be held against the wall whilst the command 3_ was entered. Control PIP's default, when a sensor is activated, is to stop, grumble, and wait for an instruction.

3.2 Valiant Roamer

This floor robot from Valiant Technology has been described as a "large Smartie", after the chocolate beans of that name. It is also a rather intelligent floor robot. More closely related to the LOGO turtle than PIP, it has a hole at its centre for a pen. Programming is via a keypad set in the top, see Fig.4.

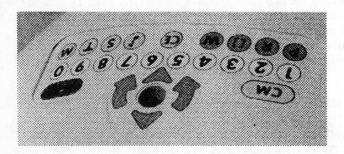

Figure 4. The Roamer keypad

3.2.1 Programming Roamer

As with PIP, the turn and move keys are symbolic. All the others, with the exception of the sound key, display the initial letter of the command, cf. one-key LOGO. The use of procedures in possible, all being sub-procedures of the inbuilt "GO" program key. An inbuilt test program is automatically run if the GO key is pressed without clearing memory (by pressing CM twice). The Roamer equivalent of the PIP program (above) is: ↑②⑤ ↓ ④⑤↑②⑤W①⑤↓②⑤ (④⑤↓②⑤ GO.
But Roamer has more capabilities than PIP. Movement and turn may be independently scaled: ↑[]②[] sets units to 2cm; and procedures defined:
P①[]R④[]↑① ↓ ④⑤[][]   defines procedure number 1 as a 45 degree corner. So, a spiral procedure might be defined: ↑[]①[]P①↑[]②[]P①↑[]③[]P①   etc.
The music command is also more powerful, the quaver symbol is followed by two numbers: length and pitch, e.g.: ♩②②⑦ . However, like PIP Roamer can only provide auditory feedback for erroneous key presses. Programming must be tested by pressing the GO button. A computer link and editing software is available.

3.2.2 Control Roamer

An optional control box may be added to Roamer to provide four two-state outputs (T), a stepper motor drive (M), and a two-state input (S). Power is available. For both input and output programming, two state logic is employed. Outputs are specifically set high or low; inputs sensitised to a high/low, low/high, or either transition. The documentation makes this clear: "High and Low refers to a voltage. Engineers use these terms because it avoids confusion caused by using words like ON and OFF". Within a Roamer program (M) and (T) commands may occur anywhere and are acted on immediately. (S) commands must always call a procedure, which interrupts the running program. Programming examples are: T①② set output 1 high; M②④⑤ , turn motor anticlockwise 45 steps; S①⑤ , sets Roamer to recognise a High to Low signal and respond by executing procedure 5. (High is coded as ②, Low is coded as ①)

3.3 Children and coding

It is clear that PIP and Roamer have radically different approaches to control. The PIP approach is simple, "Show me and tell me what to do about it.", whilst Roamer directs pupils to conceptualise the underlying electronics and logic. Though apparently similar and both using number variables, these two floor robots are positioned at very different points on the language cone. Control PIP operates at the iconic level and above, recording actions to which it will react. Roamer is operates with codes and below, right down to bits. It we place these requirements in the general context of language learning we see that PIP may be used by younger children without much fear of confusion. Control Roamer demands a much higher level of cognitive development before its programming may be understood. Indeed, some delegates at the workshop expressed doubts over the accessibility of Roamer's undoubted intellectual rigour through the coding system adopted. Has its complexity, perhaps, exceeded the capabilities of a simple input pad?

4 Models of Representation

The final group of workshops were concerned with the representation of control, both single and multi-process, at screen level and with interfaced hardware models.

4.1 Marta the Screen Robot

Eduardo Calabrese's screen robot is a LOGO microworld which can be used as a 'soft' introduction to programming from pre-school age to adults. Marta is a little screen robot which moves in a section of a "city" (Marta's world) made of (horizontal) streets and (vertical) avenues (Fig 5).

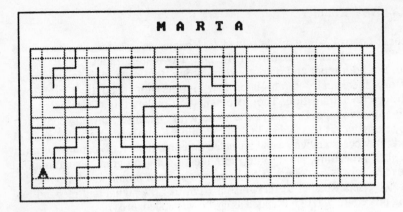

Figure 5. Marta inside a maze

She can only move forward from one intersection to the next or rotate 90 degrees left or right. At an intersection she can deposit up to 4 objects (*bips*) or pick up an object which is already there. There can be obstacles in Marta's world. Beside the border, there can be walls between intersections. Marta cannot pass through these walls. Fortunately, our robot has sensors which allow her to detect the presence of a bip at an intersection or the presence of a wall in front of her, at her left or at her right. Marta can also place or remove walls in front of her!

4.1.1 Programming Marta

Marta can be driven by single keystrokes or can be programmed by using single or multiple-line programs or by defining LOGO procedures. With these procedures Marta can be programmed both to construct and explore mazes and to carry out actions within them. A simple Marta program might be: HOME FORWARD IF CLEARRIGHT? [RIGHT FORWARD] IF HOWMANY < 2 [PUTBIP] [GETBIP].

4.1.2 Levels of representation

Referring again to the language cone, we see that two levels of representation are in operation. The screen provides a graphic representation of a computer language (character/word level) input. Compared with a floor robot, the representational level requires greater pre-knowledge of spatial and symbolic representation but the input is closer to natural language. Additionally, the language has greater power because conditionals which require an answer to a question are possible. An overlay keyboard overlay input "dictionary" would reduce the language learning requirement.

4.2 Andros

Jean-Baptiste La Palme's workshop demonstrated the parallel programming environment with which his presentation (this volume) was concerned. Written in FORTH, Andros is a simplification of an earlier language Androide. The environment under control from Andros was an electric train on a track with a level (highway) crossing and detectors (microswitches) which registered the train's passing.

4.2.1 Representing programming in Andros

The user representation of the process used the computer screen and keyboard (or screen, icons and mouse). The system demonstrated had two outputs and six inputs. The former were assigned to the letters Q & W and A & S to activate and deactivate. The input commands were two state and assigned to keys 1, 2, 3, & 4; + or - preceding the number depending upon the active transition, on or off. Other keys have additional roles. The screen display monitors the state of the events within the system, using the letters of the keyboard keys by highlighting those currently active. It will be seen that the level of both display and input is that of the code. Keys chosen, hence the letters, carry no meaning and stand as arbitrary symbols. The representation is a formalism closer to mathematics than natural language. Hence, it is straightforward to replace the letters with icons. As with the intellectual rigour of Control Roamer we need to ask what sort of mental model younger children need to possess, which they can bring to this formalism. In its present form, Andros provides an elegant formal research tool for the study of children's concepts of parallelism.

4.3 Representing Concurrent Control Programming

The question addressed in David's Argles demonstration was akin to that addressed in Andros, "How might parallel programming best be represented to the user?". Retaining his earlier approach to sequential control - a LOGO extension developed from his control program Javelin - he used the speed of the Acorn Archimedes to provide a concurrent programming environment. Separate processes were written in separate windows. Thus, the representation of parallelism was at once language and graphics based. The parallel processes represented as separate sets of instructions displayed together. It was envisaged that additional primitives might be required to talk and listen to other windows.

4.3.1 Parallel programming metaphors

In the discussion which ensued the general question of concurrency (parallelism) was aired. It seemed that a good metaphor for the requirements might be a play script. This was a (sequential language based) technique which had been developed to solve just this problem. The processors might best be considered players in a scene. Both the scene (context) and player's reactions to it and each other may be written in a sequential script. The notion of an actor waiting for a cue might be a good protocol for communication. This would imply that processes had identities which could be detected by others.

We note, that the level of representation used for this presentation was that of the word displayed graphically using windows - a separate window for each player's script.

4.3.2 Parallel programming standards group

A group was set up at the meeting to study some of these questions and consider standards. Duncan Louttit offered to coordinate this. A meeting was held shortly afterwards.

4.4 LEGO Dacta Control Lab

Ole Moller's demonstration of LEGO® Dacta's new control software and hardware required some caution in its assessment. This was because this first (as yet unreleased) product was designed for the secondary school age range. However, there was to be a primary school version later which would have a similar base. The system itself is described elsewhere in this volume.

4.4.1 Hardware

The hardware is a development of earlier Technic LEGO®, adding a temperature and rotation sensors and a sound output device. All these devices, and more, have already been used effectively in control work. The newer active LEGO® bricks are, however, quite sophisticated. The Interface improved on the earlier one: 8 outputs and 8 inputs are now provided, all the inputs had analogue capability and 4 were powered.

4.4.2 Software

The software was a radical development of LCSI LEGO® TC LOGO. Using the Macintosh interface as a basis, the new graphic user interface introduces elements of "Boxer". The "page" metaphor of LogoWriter is retained but now you can open boxes on the page. These boxes could contain: text, a picture, a graph, or monitor an input. Buttons and switches, and slider controllers could now be placed on the screen. In addition to these report pages, there is a setup and a procedures page. The command centre, for direct input of LOGO instructions, is retained. The LOGO language has been extended to handle these new entities.

Important developments include (see Fig. 6):
The setup page, through which users tell the computer how they have connected the active elements of the model to the interface.
A leap up the technological language cone from the bit to graphical representation.

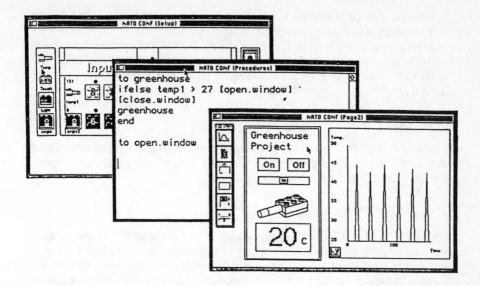

Figure 6. LEGO Dacta Control Lab pages

4.4.3 LOGO language development

Consider the following from Logotron's BBC Control Logo: IF SENSE? 5 [TURNON 3], which is described in the manual as, "if bit 5 of the input port is true, turn on bit 3 of the output port". The "ports" have now gone, replaced by messages sent and received. And these messages are translated, by the computer, into a human readable form. The interface box connections are numbered and lettered (cf. Andros) sequentially but cannot be used directly. A LEGO® element must be associated with the port. Thus, a motor on connection E becomes "motore"; a temperature sensor on connection 3 becomes "temp3". So, even at the lowest level, the equivalent of the Logotron instruction is: if temp3 > 25 [talkto "motore on], which means if temperature sensor connected to 3 registers more than 25 degrees Celsius switch on the motor connected to E. This is far closer to children's (and teacher's) everyday knowledge than was the Control Logo of a decade ago.

4.4.4 Levels of representation

Coupled with this are multiple levels of representation.
- The LOGO computer language may be used to make things happen under program control: talkto "motore on;
- clicking buttons above motore on the illustration of the Interface on the Setup page will make it rotate (clockwise or anticlockwise); and

- a button may be placed on a Report page and commands, e.g. talkto "motore on, assigned to it using a dialogue box.
- Similarly, a sensor may be interrogated linguistically: show temp3; its current value may be read above its connector on the Interface illustration; and
- its value may be displayed on the Report pages in a monitor box, on a graph in a graph box; or inserted into a text box.

4.4.5 Parallel processing

A form of "Parallel" processing is supported. Individual processes may be "launched" independently. There is also the conceptually dubious primitive "forever" which keeps a process running.

4.4.6 Discussion

If we look at LEGO DACTA Control Lab from the viewpoint of the language cone, we see that professional computer programmers have employed computer language to great effect in providing a wide range of representation above the level of code; including that of the LOGO computer language itself. The full range of representation is available: from the concrete LEGO® itself, through pictures of it on the screen, iconic buttons to control models, graphical representation of processes, and facilities to write about what is happening. Though the power of these facilities vary - there are no drawing tools to sketch prototypes, just "lineto" and "moveto" primitives - they are all present and can therefore develop. Underlying this is LOGO; and thereby access to computer language - language "beyond the byte" - which makes computer control possible.

If there is any area of difficulty which begins to appear as a result of this LEGO® development it is with LOGO itself. Early implementations running on small computers made excessive use of abbreviations, e.g. cs for clearscreen. These still abound and inhibit readability. Similarly, the prefix "set" is rather pervasive. This lexical aspect requires care if a consistent and comprehensible vocabulary set is to develop.

5 Conclusion

We have viewed the workshops from the viewpoint of how the presentations represented control. It is clear that graphic representation, i.e. illustration, of control systems poses a significant pedagogical problem; floor robots being particularly problematic. Parallel processes and the notion of independent actors, known to one another, is another potential field for development. It was also clear that the language mode of representation was distinct. It is, ultimately, the most powerful representation but entails a learning overhead. Equivalent operations in screen environments and the use of an overlay keyboard can assist the learning process.

References

Argles D. (1992), Concurrent control for children, This volume.

Calabrese E. (1991), Marta - A 'soft' introduction to programming for children of all ages, In Calabrese E (Ed) Proceedings of the Third European Logo Conference, pp. 725-30, Parma: ASI.

Doyle M.P. (1986), The microcomputer as an educational medium, Unpublished MPhil Thesis, Manchester University, UK.

Doyle M.P. (1991), Logo: once more from the top, In Calabrese E. (Ed), Proceedings of the Third European Logo Conference, pp. 1-42, Parma: ASI.

La Palme J.-B. & Bélanger M. (1992), Learning mode in the exploration of parallelism in pedagogical robotics, This volume.

Meredith M. D. & Briggs (1982), Bigtrak Plus, London: National Council for Educational Technology.

Moeller O. (1992), Hands-on control Ttchnology with new LEGO Dacta tools, This volume; LEGO Dacta, DK-7190 Billund, Denmark.

Swallow Systems Limited, 32 High Street, HP11 2AQ High Wycombe, Bucks, UK. Tel +44 494 813471

The Concept Keyboard Co., Moorside Road, Winnall, SO23 7RX Winchester, UK. Fax +44 962 841657, Tel +44 962 843322

Valiant Technology Limited, Myrtle House, 69 Salcott Road, London SW11 6DQ, UK. Fax +44 71 924 1892, Tel +44 71 924 2366

Hands-on Control Technology with New LEGO Dacta Tools

Ole Møller

Projectmanager LEGO Dacta A/S, DK-7190 Billund, Denmark

Abstract. This paper will focus on two new LEGO Dacta Control
environments:
- First Control and
- LEGO DACTA Control Lab.

Both of them build on a hands-on learning and create an environment which
encourages the student to test and observe scientific phenomena in situations they
are in control with.

Papert says: "When people fall in love with a certain type of knowledge, they
learn it very easily". Both Control environments - First Control and Control Lab
Secondary - create an environment where the student can fall in love with both the
object and the process of learning.

Keywords. Control Technology, Design, Human Interface, Learning Mode,
Programming, Project Driven Learning, Pedagogical Robotics, Sensors.

1 Introduction

Our materials always keep two principles in focus:
- The materials must provide learners with an environment that is enjoyable to
  explore.
- The materials must serve as couriers of powerful ideas.

2 LEGO Dacta First Control

First Control is a new computer independent control material. The literature has
been developed in the classroom with children and teachers who were unfamiliar
with Control Technology.

The Control Centre and the activities in the pack provide an introduction to Control Technology enabling the children to work from the first hand practical experiences in a cross curricular way.

The Control Centre can be taught to remember a sequence of instructions which it will then carry out - all keypresses will be remembered.

The Control Centre is introduced via three sets of activity cards:

- Green cards introduce the Control Centre and the electric parts.
- Blue cards to explore the main models in the pack.
- Red cards introduce problem solving activities.

Using the cards the children will be encouraged to develop their own problem solving skills and strategies by following the Technology Curriculum recommended process.

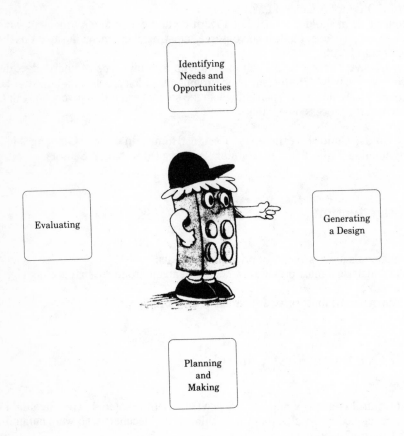

Figure 1. The problem solving process

2.1 Bricky Treads the Boards

Controlling a revolving stage in a theatre is not an everyday event for children.
   The activity cards bring the children to the fantasy world of the theatre.
   They are asked to come up with a program for the stage. Using the Program
Sheet the children get the first introduction writing a sequence of instructions.

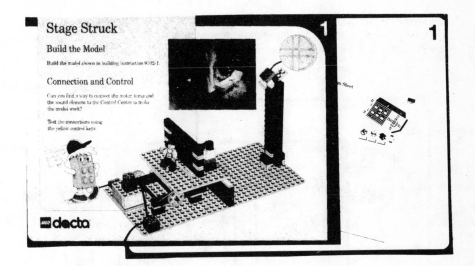

Figure 2. Student activity cards

3 Control LAB Secondary

The new Control LAB material consist of :
   - a 8 in / out serial interface
   - a Technology set. This set contains simple and more complex models to build
      dealing with the topics Communication, Manufacturing, Transportation and
      Enviromental studies.
   - LEGO Dacta curriculum materials

In the following I will mainly focus on the new Control LAB software.
Control Lab Secondary software is a computer environment built up as an
interactive book where you can:

- write procedures to control the movement of a model
- collect and display feedback from the various sensors
- create and print reports integrating pictures, text and procedures
- create interactive computer presentations for class reports and projects using active elements such as buttons and reporters.

The software is based on the LCSI LOGO. Using LOGO new tools can be created and there are full logocontrol of all the screen objects.

A Control Lab project consists of:

- One SETUP page for setting up the software with the interface box.
- One procedure page where programs are written.
- Project pages for running an experiment and writing a report.

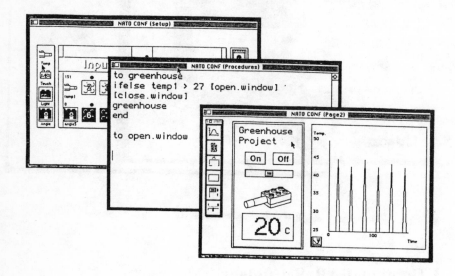

Figure 3. Control LAB user interfaces

3.1 Tomato Control - a Greenhouse Project

Computer control of a greenhouse is a stimulating and exciting project which encourages students to use the computer to solve a real life problem.

Build, connect and test

Step one is to build the Greenhouse. The Greenhouse is one of the seven main models in the LEGO Dacta Technology set.

Step two is to connect the model to the interface and do the same on the software SETUP mode.

SETUP mode is used to designate the type of sensors to the input ports, activate motors, lamps and sound elements using the mouse, visualize the values reported by the sensors. In this way the student can experiment with the model.

The temperature sensor can be calibrated and the temperature can be measured in Celcius or in Fahrenheit.

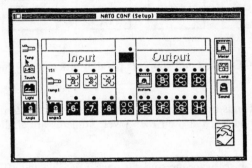

Figure 4. SETUP screen

3.2 Create a Greenhouse Control Panel

With Control Lab you can design your own user surface. The teacher can use the Control Lab software as a "development tool" for designing the user surface that goes well with the teachers methodology.

The students can by means of 6 tools (graph, text, pictures, buttons, monitors and sliders) build up their own user interfaces that the model that control the model they experiment with.

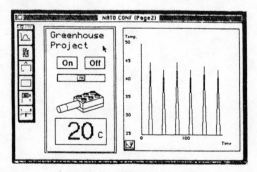

Figure 5. Project page

The tools or the screen objects are as shown below:

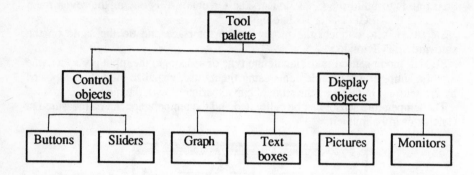

Control Lab is LOGO based and supports parallel processes, it means that more processes/models can be controlled at the same time.

With access to the hands-on tools and the powerful screen objects students are free to explore, create and learn at their own level. Students are challenged by their imaginations.

With the Control Lab software the students have a flexible and efficient tool with a wide range of applicability. The Control Lab software puts the students in focus and helps the students to learn by experimenting.

3. Tools Developed for Control Technology

3.2 Future Trends

A New Development of Control Technology

Pascal Leroux

Laboratoire d'Informatique Université du Maine, Avenue Olivier Messiaen, B.P. 535, F72017 Le Mans Cedex, France

Abstract. Our activities in control technology are based on micro-robotics environments. This paper addresses the limits due to the environment that produce "oversolicitation" of the teacher. This is the reason why we have developped ROBOTEACH, a new learning software tool which is introduced in our micro-robotics environment and which is designed to allow basic notions learning in robot driving. ROBOTEACH is a flexible learning software tool which allows the creation of a personalized learning session by the teacher. For a student the pedagogical session corresponds to a programmed access to different learning units. Thanks to ROBOTEACH the teacher manages the activities of the different students in a classroom and manages his/her time between the students.

Keywords. Adult learning, Control technology, Learning environment, Pedagogical session Generator, Programming, Pedagogical robotics.

1 Introduction

Training is becoming more and more individualized, in an attempt to improve learning and so to improve training processes. We know too that students and pupils cannot learn by themselves with only learning software; Artificial Intelligence techniques can help but still need improvements. A few learning software packages do exist but they usually are badly or not used at all by the teachers. A reason is that *this software does not take care of the teacher's role*. Most teachers think that these systems are made to replace them. It is neither true nor possible since the pedagogical situations with these systems are too limited. Another parameter of the learning software integration is their content. The more the learning software is adapted to the learner's needs and is easily usable, the more the student could learn efficiently.

In short the teacher plays a mediator role between the learner and the learning software. The teacher must *choose* the learning software taking into consideration training goals and student's level, *master* it and *adapt* the learning environment to a good use of the learning software. If (s)he realizes this integration then (s)he increases the quality of his/her teaching and therefore his/her students' motivation. The difficulty for the teacher lies in the creation of learning environments. A solution to that problem is the development of learning softwares which would be easily integrable in the learning environments. Studying the use scenarios of learning softwares can modify the design of the learning software (Vivet 1990 a).

These softwares would *help the student to learn* during the pedagogical activities and *help the teacher to manage* these activities. Programming the learning software by the teacher is a possibility to take into account the teacher's role in the learning softwares.

Our activities in control technology are based on micro-robotics environments. These environments are very rich and the students are motivated. But we meet limits of our environments in the form of an *"oversolicitation" of the teacher*. It is the reason why we develop ROBOTEACH a new learning software which is introduced in the micro-robotics environment. ROBOTEACH is designed to allow learning basic notions in robot driving. It is composed of a few learning units which will be described in another chapter. For a student the pedagogical session corresponds to a programmed access to different units. The unit planning is realized by the teacher from planning help tools. The teacher can also program his/her personal interventions during the session. Thus (s)he manages the different students' activities in a classroom and manages his/her time between the students.

ROBOTEACH is a new development of control technology. It can be seen under different points of view :
- for the student, it is a *learning software* which helps him/her during the pedagogical activities,
- for the teacher, it is a *pedagogical session generator* in control technology which facilitates his/her activity management.

In this paper, we will describe above all the pedagogical session generator aspect.

2 Remarks about ROBOTEACH development

ROBOTEACH is more than a software designed in a computer science laboratory. It is developped as a part of a program which links the Ministry of Research and Technology, the "Pôle Productique" of the town of Saint-Nazaire (France) which is a center of innovation and transfer of technology, our computer science laboratory and a professionnal training center of Saint-Nazaire (A.F.P.A.). ROBOTEACH will be used by schools, high-schools, professional training centers and companies.

ROBOTEACH is created above all from observations coming from our process of learning experiments at low qualified adults' training level (Vivet and Parmentier 1991). These activities can easily be adapted in elementary schools. Besides the development of our last activites, early works have been realized by Bruneau in classroom (Bruneau 1989, Vivet and Bruneau and Parmentier 1990 b). Some reflexions about professionnal training are interesting for elementary schools and for control technology. The evolutions and the attempts in all trainings (even the fruitless attempts) and in all sciences (didactic science...) must be taken into consideration to improve learning based on control technology environments. Otherwise we would create closed environment and the teaching would become less effective. J.W. Clancey claims how desapointed he is from this behaviour in the development of Guidon-Manage (Clancey 1992). Nevertheless we must remain simple and not dream to a too wonderful system. We must not mimie Icare's fable and burn our wings at the sun (Blandow 1992).

3 Description and limits of micro-robotics environment

Our control technology activities are built from micro-robots based micro-worlds (Vivet 1986). The concept of micro-world has been described by Papert (Papert 1980). The laboratory has developped an interface and LOGO primitive commands to activate our micro-robots with a computer. The micro-robots are real objects: the students can touch, move, build, modify and activate them. Thus the students can understand better what we want to teach them and build their own knowledge. The micro-robots play the same role as Seymour Papert's gears which have facilitated the introduction of mathematics in his mind (Papert 1980). The advantage of technology learning is that lessons can be based on devices. It is the reason why the control technology environments are effective: they are often based on the link between computer and device. Among these environments we think of Nonnon's experimental environments (Nonnon and Laurencelle 1984, Nonnon 1986) in which the pupils define the parameters and the starting of the experiments. Deleting the devices in the control technology environment is more or less deleting the control technology. For technology learners, the devices are more than a help in their learning process and a media embeding relevant knowledge ; *the devices are an essential mean of communication between the learners and the teachers.*

In our activities we want to introduce the basic notions in technology. Our methodology imposes that before every pedagogical activity we clarify the goals in terms of contents, knowledge abilities to be developped (for example facilitating acquisition of basic knowledge in technology), the kind of abilities that we want to develop (for example organisation/planification of a set of events) and the pedagogical attitudes (for example robot design). A synthesis of our approach is available in (Vivet 1989 a, Vivet 1989 b, Vivet and Parmentier 1991).

The micro-robotics activities are very rich and the amount of communications between the teacher and the learners becomes too important. This "oversolicitation" of the teacher has an effect on his/her teaching: it is less effective. The teacher is "oversolicited" because (s)he plays four different roles which are :

- *transmission* of information, knowledge, exercices with the help of reading sheets,
- *teaching programming concepts* in order to help the students program the movements of their micro-robots,
- *help the learner to activate* the micro-robots and *to debug* the command programs,
- *managing the activity* to enable the student to discover new concepts, i.e. managing the pedagogical interventions during the activities.

At present to decrease the "oversolicitation" of the teacher in a low qualified adults context we are often 2 teachers for 9 learners. This wonderful pedagogical situation is also luxury and very expensive. So, large scale dissemination is not possible and it would very difficult to adapt our training attitude in a classroom of 15 students and 1 teacher.

It is to improve the task of the teacher and improve the learner's autonomy that we begin to develop ROBOTEACH. ROBOTEACH is realized to allow basic notions learning in robot driving such as:

- physical composition of a micro-robot,
- basic notions in technology, automatic production process,
- links between the computer, the interface and the micro-robot,
- driving concepts.

It is integrated in our micro-robotics learning environment taking in account the role of the teacher.

Now we are going to explain how ROBOTEACH helps the teacher in his/her four roles to decrease his/her "oversolicitation".

4 Electronic cards

During our training process we intervene very often to provide knowledge, pieces of information and exercices with the help of reading sheets. For a given (or specific) activity (building a micro-robot for example) we provide the same information for all students. So it is very easy to put together these pieces of information in a base of electronic cards: we implement this solution in ROBOTEACH to decrease document distribution by the teacher.

Electronic books and few Computer Assisted Learning systems are built with electronic cards. They are more attractive than the traditional books because the sound, the picture, the text are mixed and the access to the electronic cards is faster. But which observations must we take into account before developping the electronic cards which will be implemented in ROBOTEACH ?

Any beginner who uses an electronic book without any computer knowledge encounters problems. In most of the electronic books, you can access any card. It is interesting for those who know the card they need but not for a neophyte. The neophyte turns the "pages" without understanding the pieces of information that (s)he reads. For him/her, the series of electronic cards are too complex. (S)he likes to use an electronic book only after understanding the links between cards to find an information. But to reach this step, *(s)he must be guided* during his/her learning activity process. Otherwise (s)he learns nothing and wastes his/her time. More important, (s)he can be disgusted by electronic books and perhaps by computer. The initial approach of a computer is essential to like working with it.

The electronic books are tools and as all tools they must be used in a good context and in an effective, well organised learning environment. For a beginner, they are not sufficient to enable him/her to learn alone. (S)he needs teachers, exercises, courses and many helps. The pedagogical work is always very important indeed *even more important* with these new learning systems. Perhaps too many teachers forget this important rule. For example, the electronic tutors, which are sold with the softwares, are made for people who already have used a computer. They are very good for the computer scientists but not for neophytes. They do not use the same language. Each domain has its jargon. Unfortunately we have met teachers who use the electronic tutors with neophytes and without adaptation of the learning environment. Everywhere to be understood *we must adapt our language* to the persons who listen to us. It is one of the first rules in education too.

Nevertheless we think that the systems based on electronic cards can be good tools. They can enrich the learning environments. The electronic cards in ROBOTEACH replace the reading sheets used in the micro-robotics environment.

Each card or group of cards shows a simple notion in technology or an exercise or an information about the working environment. Our goal is not to realize an electronic book but to have many simple electronic cards. Thus the teacher can plan the presentation of the notions as (s)he wants. The learner has the document (s)he needs at the best pedagogical moment. Perhaps the teacher will give a free access to electronic cards or still use the reading sheets. (S)he will make his/her didactic choices. Before (s)he programs ROBOTEACH in fonction of his/her choices. More information about that is available in the chapter 7. The electronic cards allow to avoid documents distribution by the teacher. If (s)he decides that the content of electronic cards are not adapted to the learners then (s)he does not used them. We are thinking of adding to ROBOTEACH a tool which helps the teacher to realize his/her own electronic cards. Thus (s)he can adapt the content of the electronic card base to his/her lessons and learners. This tool will not be described in this paper.

5 Activating programs generated from a micro-robot description

In our micro-robotics environment the major problems are met when begining a training session with neophytes in computer science. To activate the micro-robots the learner must program the driving commands and so use a programming language (LOGO). It is described as a very simple programming language but the learners face many problems to realize good commands and to activate the micro-robots. So the teacher must frequently intervene to give explanations about LOGO. We must keep in mind that teaching a programming language is not our goal at this stage. In such a context the learners acquire two types of knowledge: basic notions in technology and basic notions in programming. It appears to be better to split difficulties and the introduction of the basic notions in technology would be sufficient at the beginning. Nevertheless once the learners are accustomed with the environment it is interesting to take the advantage of the micro-robotics environment to teach programming.

To focus the neophyte's attention on the technologic notions and to delete the interventions about programming, we have introduced an *expert system which generates the driving LOGO programs from a description of the micro-robot built by the learner* (Leroux and Bruneau 1991). The description language allows a description dealing with the morphology, the kinematics and the dynamics (informations to activate) of the micro-robot. In a first step, the learner gives a description of the robot. Then (s)he can use the generated driving programs to activate his/her micro-robot : thus (s)he gets a feedback from the given description. If a bug appears (difference between what is expected and what is obtained) it is necessary to improve the description. No correction is done at the level of programming language.

The expert system's goal is to acquire from a discussion with the learner a rational description of the device. The learner's goal is to activate his/her micro-robot. They cooperate with each other for their own goals. Several cooperation processes between the learner and the expert system are interesting: exploring micro-robotics micro-world, interacting to describe and to understand the micro-robot working, activating the micro-robot and finding the description bugs (Leroux 1992).

A first release of this expert system works for very simple micro-robots, but we have not yet tried it in real training environment. Improvements are necessary to facilitate its use by the learners. This improved release is integrated in ROBOTEACH. To help the explanation about the pedagogical situations, we are going to give more information about characteristics of micro-robot.

A micro-robot is composed of bricks. Some of these bricks are motors. When you activate a motor the micro-robot moves. You obtain a movement of a part of the micro-robot or of the whole micro-robot. Diagram 1 shows the kinematics chain of the micro-robot.

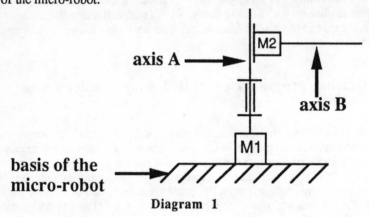

Diagram 1

If you activate the motor M1, the whole micro-robot moves. If you activate the motor M2, only axis B moves. For the rest of this paper :

- an *axis* will be a part of the micro-robot which is moved by a motor,
- a *kinematics chain* will be the link between several axes. In the diagram above, axis B is linked to axis A because when you activate M1 these axes A and B move whereas when you activate M2 only axis B moves.

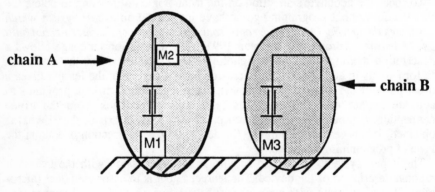

Diagram 2

A micro-robot is composed of one or several kinematics chains. In diagram 1, the micro-robot is composed of one kinematics chain while in diagram 2 there are two kinematics chains. The movements of chain A and chain B are independant but we must synchronize the movements between these chains to realize a good activation of the micro-robot. In our pedagogical activities the micro-robots are classified in two categories: *one or several kinematics chains*.

During the activities of description and the activation with ROBOTEACH a didactical situation depends on :
- the category of the micro-robot described,
- the describing, or the activating way.

A beginner describes and activates more easily his/her micro-robot in an "axis after axis" way. Once the learner is accustomed (s)he can describe all axes before activating the micro-robot. A learner, who is accustomed to describe and to activate a micro-robot which has one kinematics chain, becomes a beginner again when (s)he must describe a micro-robot with several kinematics chains. Therefore in each micro-robot category we define two levels of learner: *beginner and confirmed*. The beginner describes and activates his/her micro-robot axis after axis. The confirmed learner describes all axes in one step then activates the micro-robot.

In the first release the learner was describing all axes before activation. The description language between the system and the learner was based on questions and answers. Now the learner has not only questions and answers, but also menus and graphic tools at his/her disposal. We adapt the description language according to the pedagogical situation. For the two categories of micro-robot the following table synthetises how is composed for us a pedagogical situation.

learner's level	kind of interaction with the system	which description which activation
beginner-1	graphic tools	description and activation of the micro-robot axis after axis
beginner-2	questions and answers menus	description and activation of the micro-robot axis after axis
confirmed	questions and answers menus	description of all axes then activation of the micro-robot

The questions and answers for beginner-2 will not be the same as for the confirmed learner. The graphic tools allow to describe the micro-robot axis after axis with geometric symbols. The graphic objects look like real objects (motor, captor...). Thus a beginner can describe more easily his/her micro-robot with these symbols. The jargon used in questions and answers can also be introduced thanks to graphic objects. It is not interesting to describe all axes with graphic tools and after to activate the micro-robot because the learner must manipulate many symbols and can waste time. A confirmed learner will describe faster all axes with questions and answers. It is the reason why we use graphic tools only for the beginner level.

We have not yet worked enough on the dialogs, menus and graphic tools. Therefore we do not give more informations about this important point. Nevertheless we can say that the description language must not be another programming language. Difficulties occur in the creation of the interactions managed by the system.

The choice of the level will be made by the teacher before the session (more information is available in section 7 hereafter) or by the system in few cases (according to the session time and the type of exercise).

Here are two examples in the case of choice made by the system :

If the learner is at beginner-2-level, after three good descriptions of micro-robots with one kinematics chain the system increases the level to confirmed learner.

If the learner is "confirmed", after five good descriptions of micro-robots with one kinematics chain the system decreases the level to beginner-1 and proposes an exercise with a micro-robot which has several kinematics chains.

6 Activation and debugging

In this section we give only the ideas of the help that ROBOTEACH will provide to the learner during activation, debugging phases.

6.1 Activation

The learner activates his/her micro-robot using the driving LOGO procedures generated by the system. Therefore activation with keys on the keyboard of each axis of the micro-robot is possible. Interactive help will be useful to create complex movements (movement which needs synchronisation of several axis movement).

6.2 Debugging

When the learner activates his/her micro-robot (s)he receives a feedback about his/her description. A dysfunction of the micro-robot leads to a modification of the description. The learner will give to the system information about the activation errors from observed bugs. The system will provide few remediation tools. Two types of tools will be useful:

- *general tools* like a program called DIAGNOSTIC which returns the static state of the interface,
- *tools associated to the kind of micro-robot* built by the learner. A data base of micro-robots descriptions with the remediation tools associated will be included in the system.

When all remediation tools are used and the problem is not solved then the system asks for the teacher's help.

7 Help to manage the pedagogical situations

Our micro-robotics environment allows us to create pedagogical situations based on *knowledge discovery*. The environments allowing knowledge discovery transmission are not enough used. And yet it is one of the best means to learn knowledge and acquire abilities. A notion is better understood and therefore acquired when discovered by the learner. Moreover the learners are more motivated.

When we deal with knowledge discovery we do not want to leave a student following any thought and waiting for an event occuring. For a good discovery we need :

- a *guide* (teacher) who just shows the goal and helps the learner only when (s)he really needs it,
- a *learning environment* with strong potential of rich discoveries and with all useful discovery tools.

The metaphor of the walk in the mountain can help us heir. The teacher is the mountain guide. The students are few persons who want to know better the mountain and its traps. The learning environment is the mountain with some tracks imposing choices. The discovery tools are the equipement, a compass, a map... The goal is to reach a magnificent place at the top of the mountain. The students and sometimes the teacher have never seen this place. At the beginning the guide gives a point in the mountain where the group must go. After (s)he follows near to the students. (S)he never is ahead the group to show the way. It is the best means in order that the group discovers the walking problems. (S)he intervenes only when there is a serious danger or when the group asks him/her some information that cannot be found by themselves. The students solve the problems by themselves, together. When they arrive at the wished place, they discover a magnificent point of view on a lake and on valleys. The learners *will never forget* this view, how to walk along mountain tracks and the guide's advices.

This metaphor is idyllic but the teachers must try to create richer learning environments. The LOGO environments give good examples of discovery learning environment notably in mathematics (Papert 1980, Vivet 1981). Control technology is a domain in which this pedagogical method is very useful. It is a great pity that control technology is too few spread in schools and particularly in the elementary schools.

Knowledge discovery is a very important point of our learning strategy. The introduction of ROBOTEACH in the micro-robotics environment must not delete this strategy. A good integration of the teacher must be done. The pedagogical sessions with ROBOTEACH are based on an alternance of discovery by assembling or design of micro-robot (knowledge discovery is made without the computer) on the one hand and knowledge discovery thanks to electronic cards or during the description, the activation of the micro-robot, or during the debugging (knowledge discovery is made with the computer) on the other hand. *A good alternation between these discovering phases and the teacher's intervention must maintain a real motivation of the learner*. Thanks to his programmed interventions, the teacher improves the follow up of the learners progression, and, from a better management of his/her own time, he/she intervenes at the best pedagogical moments.

In ROBOTEACH the teacher plans the pedagogical sessions for each learner. A session is composed of interventions made by the teacher, self execution of different units (description, activation, debugging, lesson (electronic card use)). The teacher specifies the chain of the units with his/her interventions like in diagram 3.

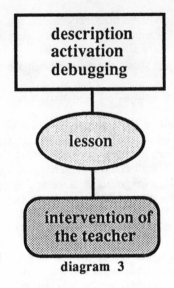

diagram 3

For each unit the teacher specifies :

- for the lesson, useful electronic cards and exercices,
- for description, activation and debugging, the exercices (building or designing a micro-robot with one or several kinematics chains), the level of the learner (beginner-1, beginner-2, confirmed) which defines the kind of interactions between the system and the learner and the way of description and activation, the knowledge electronic cards useful and the remediation tools available.

8 Conclusion

One of the major problems with the Computer Assisted Learning lies in software use. The students do not use them because of a lack of knowledge or abilities, they get lost with no relevant help. The teachers use badly or do not use them because learning softwares are not enough flexible and not easily adaptable in teaching environment. A design from a correct and complete specification of the scenario of use is necessary.

The teaching softwares must evolve in several ways :

- developping knowledge discovery strategies,
- being parameterised by the teacher to facilitate their integration in learning environments,
- linking to real world to increase the learner's interest.

In short, these new softwares will have to help the learners and the teachers and we think that control technology is a favorable domain to create such tools.

We hope to integrate these ideas in our new development (ROBOTEACH) in control technology. It will help the learner to discover the basic notions in technology through the use of electronic cards and through the activation of real micro-robots. It will help the teacher managing the teaching space thanks to the programming of the pedagogical sessions for each learner. ROBOTEACH is a flexible learning software which allows the creation of personalized learning session.

The first use of ROBOTEACH in real training will give us many answers about our ideas on the evolutions of learning softwares.

References

Blandow, D. (1992) Tools to overcome thought barriers - A new element of effective technology education. First International Conference on Technology Education, Weimar, Germany, April 25/30, 1992

Bruneau, J. (1989) Remarques autour d'une activité de robotique en classe de 6ème. Actes du Premier Congrès Francophone de Robotique Pédagogique, Le Mans, 24-27 août 1989

Clancey, W.J. (1992) Guidon-Manage Revisited: A Socio-Technical Systems Approach. Second International Conference on Intelligent Tutoring Systems, ITS '92, Montréal, Canada, June 10/12 1992, Lectures Notes in Computer Science, Vol. 608 : Springer

Leroux, P., Bruneau, J. (1991) Coopération entre un élève, un environnement de micro-robotique et un système expert de pilotage de micro-robots. 3ème Congrès Francophone sur la Robotique Pédagogique, México, 21/23 août 1991

Leroux, P. (1992) Cooperation between pupil and expert system to drive a micro-robot. First International Conference on Technology Education, Weimar, Germany, 25-30 April 1992

Nonnon, P., Laurencelle, L. (1984) L'appariteur robot et la pédagogie des disciplines expérimentales. SPECTRE, pp. 34/36

Nonnon, P. (1986) Laboratoire d'initiation aux sciences assisté par ordinateur. Publication Vice-décanat à la recherche, Faculté des sciences de l'éducation, Université de Montréal

Papert, S. (1980) Mindstorms, Children, Computers and Powerful Ideas. New York: Basic Books 1980

Vivet, M. (1981) Apprentissage autonome sur un usage de la technologie informatique dans l'éducation. annexe du rapport SIMON "Éducation et informatisation de la société", la documentation française

Vivet, M. (1986) Pilotage de micro-robots sous LOGO: un outil pour sensibiliser les personnels de l'industrie à la robotique. 5ième symposium canadien sur la technologie pédagogique, Ottawa, 5-7 mai 1986, disponible dans le livre: A l'école des robots" diffusé par la robothèque du CESTA, p.195-210

Vivet, M. (1989 a) Which goals, which pedagogical attitudes with micro-robots in a classroom ?. NATO, Workshop on Advanced Educational Technologie, "Student development of physics concepts : the role of educational technology", Pavia, Italy, October 4/7 1989

Vivet, M. Micro-robots as a source of motivation for geometry. In: J.M. Laborde (ed.), Intelligent Learning Environments: The Case of Geometry. NATO ASI Series F, Vol. 117. Berlin: Springer-Verlag 1993

Vivet, M. Uses of ITS: which role for the teacher. In: E. Costa (ed.), New Directions for Intelligent Tutoring Systems. NATO ASI Series F, Vol. 91. Berlin: Springer-Verlag 1992

Vivet, M., Bruneau, J., Parmentier, C. Learning with micro-robotics activities. NATO, In: M. Hacker, A. Gordon, M. de Vries (eds.), Integrating Advanced Educational Technology into Technology Education. NATO ASI Series F, Vol. 78. Berlin: Springer-Verlag 1991

Vivet, M., Parmentier, C. (1991) Low qualified adults in computer integrated enterprise: an example of in service training. IFIP TC3/WG3.4, Alesund, Norway, 1-5 July 1991, in "TRAINING : from Computer Aided Design to Computer Integrated Enterprise", B.Z. Barta and H. Haugen (eds.), North-Holland 1991, 261-272

A Data Acquisition System in a Learning Environment

Pierre Nonnon and Claude Johnson

Laboratoire de Robotique Pédagogique, Université De Montréal, CP 6128
Montréal H3C3J7, Québec, Canada.

Abstract. This is version 1 of a data acquisition system presently under development at our laboratory. It is meant to be used in a learning environment. It is different from industrial data acquisition systems. It allows one to put the emphasis on pedagogical objectives. For example, it gives a much more detailed and explicit explanation of the concept of transduction.

This can be achieved by giving the student an opportunity to design and develop his/her own measuring instruments and by giving him/her access to graphical and mathematical tools which will make the transfer from physical to mathematical variables much more understandable .

Keywords. Assimilation, Cognitive tool, Control technology, Data acquisition, Representation.

1 Introduction

Physical phenomena which are being measured are rarely interpreted directly by a computer. They must undergo a number of transformations before a computer can analyse them.

To give an easy example, the temperature one can read on a thermometer is a causal interpretation of the expansion of a liquid such as mercury or alcohol.

In a data acquisition process, the visual reading of a temperature can not be easily used; one usually prefers to use a thermistor, i.e. a resistance which varies according to the temperature. Temperature causes this resistance to vary and the resistance, inserted in a bridge, causes the voltage to vary; the voltage is transformed by an A\D converter into a variable with a numeric value; this value is then analysed and represented by the computer (see diagram below).

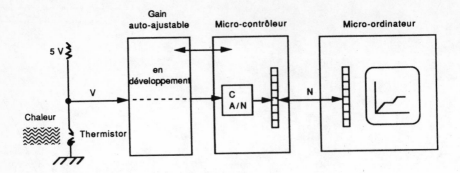

Fig. 1. Data acquisition process

Whenever a physical phenomenon undergoes a transformation, we know students are faced with the problem of understanding and interpreting that transformation. It would be nice if they had already mastered the graphical and mathematical tools which facilitate the understanding and interpretation process of the transformational process.

Mastering these tools can be done in a concrete, multisensory environment, one in which the student can not only observe, manipulate and understand the interaction between variables but also watch a dynamic, real-time representation of the interaction at hand.

We have developped such a pedagogical environment, a laboratory for computer assisted Sciences. We have described it with the concept of "cognitive lens" or "cognitive spectacles" (Nonnon, 1986).

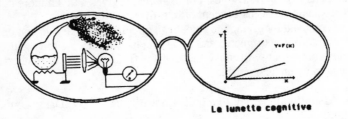

Fig. 2. Cognitive Spectacles

In the "cognitive lens/spectactles" context, the main objective is to give students the opportunity to understand and manipulate a number of graphical and mathematical cognitive tools.
In the "Oscilloscope" context, which we are currently developping, the main objective is to give students an opportunity to understand and interpret the transformational process a variable undergoes in a data acquisition system.

The "Cognitive lens" context, then, is prerequisite to the Oscilloscope context.
The first one will offer the students cognitive tools which are necessary for the understanding of the other.

The challenge is not to interpret mathematical representation while one percieves current physical phenomena, but to understand non perceptible physical phenomena with the help of cognitive tools such as graphical representations and imathematical equations.

The idea is not to go from concrete to abstract (to assimilate the abstraction via the concrete phenomenon), but to go from a non perceptible abstraction to an abstract phenomenon (to go from abstract to abstract). In order to perform this substitution, a student must be able to manipulate the cognitive tools which are offered to him/her and the presence of the skill must of course be established.

2 A description of the equipment

Our system is called Oscilloscope. The Macintosh version of version 1 of Oscilloscope requires an asynchronous micro-controller built around the 8052 AH Basic chip made by Intel. This controler has a ROM Basic which can handle interupts.

Our system adopts the "look and feel" of all other Macintosh products. Under the Multifinder, it can operate in background mode. This can be used when experiments take up a lot of time, such as when one studies the performance of a solar oven or the growth of a number of plants.

The *first time* one uses this system, a personal calibration file must be created; this is achieved by opening the "File" menu and selecting "New". This file will store calibration values for up to 8 channels and can be saved on diskette.

If one already has a personal calibration file, one will simply use "Open" in the "File" menu and select the desired file.

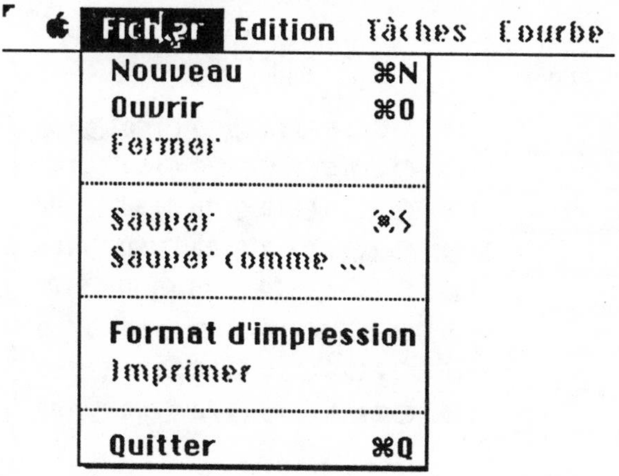

Fig. 3. File menu

3 Channel identification

Each channel can be used to measure a pressure, a temperature... Channels can be named Manometer, Thermometer etc... This name tag can be changed at all times, by selecting "Identification and Gain" in the "Task" menu.

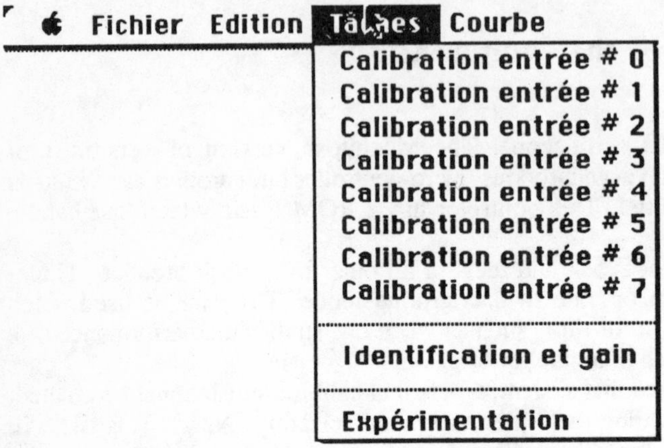

Fig. 4. Tasks menu

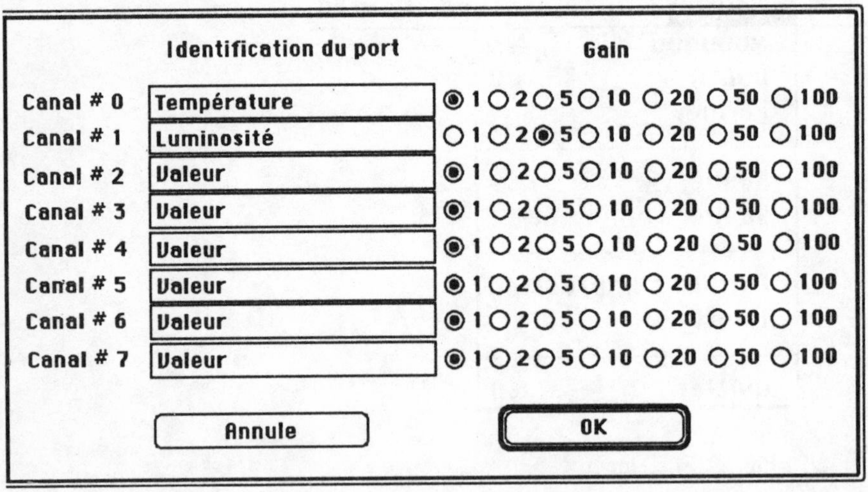

Fig. 5. Programmable gain and channels identification

4 Programmable gain

On all channel, the gain of all amplifiers is preset to a default value. Depending on the order of magnitude in differences of potential (voltage) observed at the input, the student will select a value for the gain, trying to cover the greatest possible distance on the graphic page. The optimal gain is the gain which gives the converters voltages between zero and five volts.

We are currently developing a auto-adjusting gain system. This will protect the input channels by accepting voltage values between one Milivolt and 2500 Volts and also make gain adjustment transparent to the students. This system will not only adjust the gain but will give the computer the value of the gain (the scale), so that the computer can convert the output data, using that scale. The tables and graphs will then be constructed according to the real value of the inputs.

We think this transparent approach will be beneficial to the students, as it will allow them to concentrate more on the design and calibration of their measuring instruments, leaving the complex technological aspects to the program.

The output values of the A\D converters must be coupled with the physical phenomenon in real time. To implement this, students are given a visual representation of the interaction between the variable of the phenomenon - for example, temperature as read on the thermometer - and the output value of the A\D converter.

The first step before conducting an experiment, then, is to calibrate and to name the channels which will be used.

5 Channel calibration

One can select the desired channel in the "Task" menu , or one can change the calibration. Two windows are then displayed. One is used to enter the data, the other to give a graphical representation of the data and offer a transfer curve.

As with every piece of Macintosh software, all you have to do is click on the desired window to bring it to the foreground.

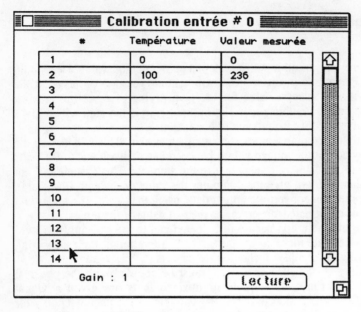

Fig. 6. Channel calibration

Steps towards calibration
1. Bring the data input window to the foreground.
2. Connect the external electrical transducer to the input channel of the converter (if you have not already done that).
3. Click inside any empty cell of the first column of the data input window and type in the real value you can see on the measuring device you are using as a standard (thermometer, voltmeter...). The values will be arranged automatically from smallest to highest at a later stage.
4. Bring your external electrical transducer to the value you typed in during step 3
   Example
   Calibrate a thermistor (an electronic themometer)
   i. Bring the water to 0 degree Celsius (add ice cubes, or use any other method).
   ii. Type in 0 for 0 degree Celsius, in the first column. Click on the "Read" button when the 0 value will be reached.
   iii. Repeat stages i and ii a number of times, for as many temperatures as you want to get. If you are certain the physical phenomenon is linear, two values will be enough.
5. Bring the graphics window to the forefront with a click. As you do this, the computer will start analysing your data, eliminating typed values for which you have not "read" the external phenomenon; it will also place the values in increasing order.
6. Points corresponding to your measures are placed on the graph, and are linked with segments of a straight line.

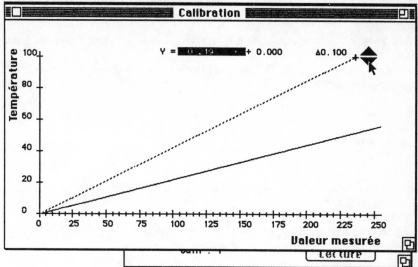

Fig. 7. Calibration graph

7. The next step is to establish the equation representing this set of points.
8. The default equation is the first degree equation coresponding to a straight line. It is displayed at the top of the graphics window with a and b parameter set to zero.

$$Y = a\,X + b$$

9. The two triangular icons to the right of the equation can be used to change the values for a and b. This will increase or decrease the value of the selected parameter by the value of delta (increment). To change a value, you first select the parameter, or delta, by clicking on it. This is pretty much the same as changing the time on the system clock available in the Apple menu.
10. Every time you change the value of a parameter, the corresponding curve or straight line is redrawn.
11. If you think the points correspond to a second degree equation, go the the "Curve" menu and select the appropriate option. In the case of a second degree equation, displaying the calculated curve is a lot slower.

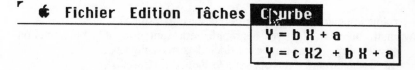

Fig. 8. Equation menu

12. Adjust the theoretical curve by changing the parameters of its equation so as to superimpose the curve over the curve representing the points of the experiment.
13. After step 12, channel calibration is considered as completed. You have found the transfer equation.
14. Repeat steps 1 to 12 for the various channels on which you wish to link a measuring instrument.

6 Experiment

When the calibration of a channel is completed, students can use that channel to conduct experiments.

Two kinds of experiments can be conducted.

1. Observing a phenomenon varying in relationship to an other phenomenon. Two calibrated channels will be necessary.

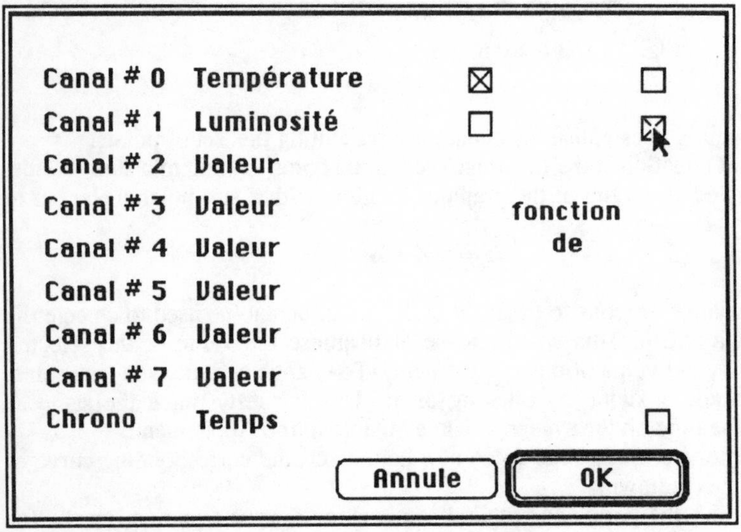

Canal # 0	Température	☒		☐
Canal # 1	Luminosité	☐		☒
Canal # 2	Valeur			
Canal # 3	Valeur		fonction	
Canal # 4	Valeur		de	
Canal # 5	Valeur			
Canal # 6	Valeur			
Canal # 7	Valeur			
Chrono	Temps			☐

Annule   OK

Fig. 9. Channels choices

2. Observing a phenomenon varying in relationship to time.
   Any instrument which gives a graphic representation of a phenomenon varying according to time can be considered as an oscilloscope.
   Click in the box to the right, opposite to the word "Chrono".
   All you have to select now is the dependant variable.
   Indicate start time and interval values between measures.
   To change a start time or an interval value you first select it with a click and then use the Up or Down arrows to set and display the desired value.

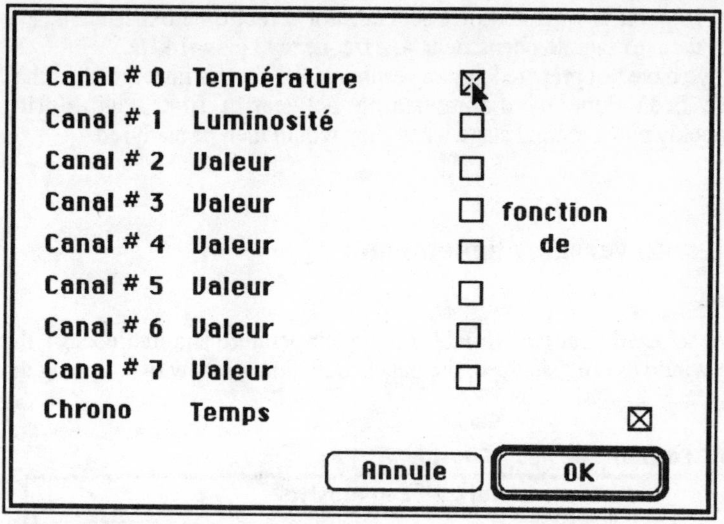

Fig. 10. Channels choices

If you want to start an experiment when an external phenomenon has reached a certain condition, you will have to select the "external event" option. The microcontroler can wait for a signal on the interup input line to reach its high level before it starts reading measures.

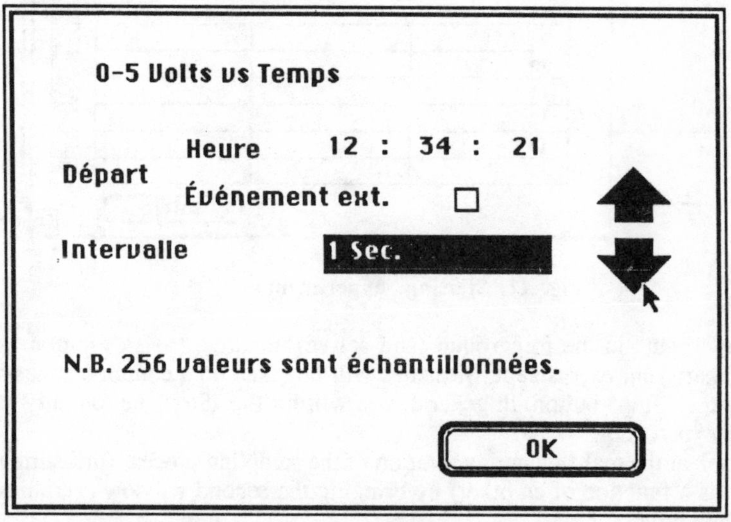

Fig. 11. External events

The minimum interval is 1/10 000 of a second. For a real time experiment, it is recommended, then, to sample phenomena at a frequency below 1 kHz.

As authors, we have not prepared any experiments for shorter intervals than that. It could have been done by distinguishing between a first stage storing conversions quickly and a second stage when they would then be analyzed.

7 Phenomenon versus phenomenon

Once you have selected your two variables (here temperature and luminosity), the two following windows are displayed; the data acquisition window overtopping the graphics one:

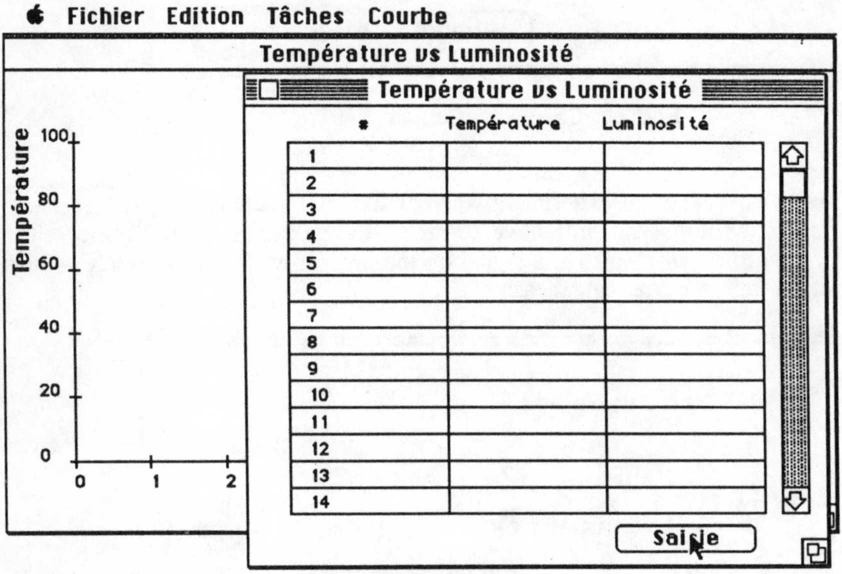

Fig. 12. Starting experiment

Click on the "Input" in the foreground (and active) window, (Saisie) button as soon as you start your experiment. Measures will be taken on a continuous basis until you hit the "Stop" button. In general, you will hit the "Stop" button only at the end of an experiment.

You can look at the real-time representation of the sampling process (measuring one channel as a function of an other) by bringing the second window (graphics one) to the foreground.

Values being displayed in the numeric window are the 0 to 255 values measured by the A\D converters.

In the graphics window, these values are transformed by the transfer equations constructed during the calibration operations.

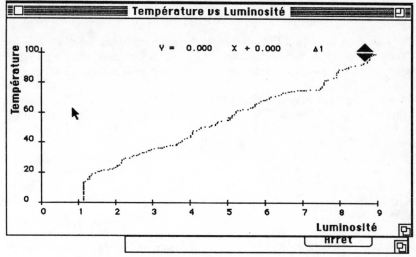

Fig. 13. Graphical results

Once the input stage is over, you can establish the equation relating these two variables by exactly the same process you used when defining the tranfer equations.

You can vary the values for the a and b parameters for a first degree equation of the $Y = a X + b$ format; or you can vary the a, b and c parameters for a second degree equation of the $Y = aX^2\ bX + c$ format.

8 Saving on diskette

Experiment stage

During this stage, only the "Save as" option is selectable. A text file is created, storing the results of the experiment.

Calibration stage

If you are not in the Experiment stage, you are in the Calibration stage. Two files are created: an X file and an X.Text file. The X file is your calibration file for the 8 channels. It is saved under a format which "Oscilloscope" can read. The X.Text file stores the same data, but in TEXT format. It can be read by any word processor or piece of software which recognizes this format, such as "Cricket Graph".

9 Using "Cricket Graph"

In order to make a model for the interactions between variables, we did not really need to integrate a great number of complex functions. However, since our data is available in TEXT format, a graphing package such as "Cricket graph" can be used.

Right after you have started "Cricket graph", select "Close" in the File menu. It will close the default window. Then select "Open" and click on "Show all TEXT files"; then open the desired TEXT file.

Refer to the "Cricket Graph" manual for all other operations.

10 Conclusion

"Oscilloscope", our data acquisition system, is meant to allow students to design and construct efficient measuring instruments such as scales, thermometers, voltmeters,etc.

Students will have to calibrate their instruments by using graphical and algebraic tools. These tools will transfer into Kilograms or degrees Celsius the values converted into digits by the A\D converter.

With this system we are also hoping to validate various pedagogical hypotheses. For example, establishing whether the representations students mastered as they studied concrete, perceptible phenomena in cinematics in our "Scientific gymnasium" environment will or will not be transfered by the students to a different domain, namely the domain of abstract electronic phenomena which cannot be percieved by our senses.

References

Le Touze, J C., Beaufils, D., (1990) Plan incliné: un logiciel d'acquision automatique et de traitement numérique de données. *Informatique et pédagogie des sciences physiques. Union des physiciens.* Paris

Marchand, D (1991) *Mise en place d'un laboratoire de sciences physiques informatisé pour une classe de 24 élèves de seconde.* in Robotique Pédagogique. Les actes du 2 congrès international, Vice-décannat à la recherche, F.S.E, Université de Montréal, Montréal Canada

Nonnon, P. (1986) *Laboratoire d'initiation aux sciences assisté par ordinateur.* Publication du vice-décanat à la recherche,Faculté des sciences de l'éducation, Université de Montréal, Montréal

Rellier, C. (1990) *Faire vivre les grandeurs électriques en régime sinusoïdal.*Paper presented at the Quatrième journées "informatique et pédagogie des sciences physiques", Toulouse, France

Stein, S.J. (1986) *The computer as lab partner: Classroom experience gleaned from one year of microcomputer-based laboratory use.* Jounal of Educational Technology Systems 15:1-15

Concurrent Control for Children

David Argles

Department of Computing, King Alfred's College, Winchester, SO22 4NR, United Kingdom

Abstract. It may be argued that control situations are inherently parallel in nature. It may also be argued that, as human beings, we inherently parallel process. Yet parallel processing is the domain of degree and post-graduate studies. Need this be so?

This paper proposes an approach for making concurrent programming accessible to young children. The claimed benefit is that the approach more closely matches the problem to be solved, making it easier to understand and the program easier to use.

Keywords. Concurrent programming, Control technology, Cooperative learning, Elementary education, Human interface, Parallelism, Problem solving, Programming, Pedagogical robotics.

1 Computers and the Real World

In this paper, I want to consider the issues of parallel programming, of encouraging modular thinking, and of graphical approaches to interaction with the children.

Right at the outset, in my imagination, I can hear a chorus of protest. Surely this is all highly esoteric, and quite divorced from current technology and the world of the child? The more I have looked at this, however, the more convinced I have become that such issues are familiar to us all in practice, that they are helpful, and that they need careful investigation. Let us start with parallel processing.

1.1 I am a parallel processing being

There is a potential minefield here. I would have liked to start by comparing myself to a parallel processing environment, where there may be several tasks being processed on one processor, or on several processors. I do not wish to be ensnared in arguments about whether I have a brain that is a multi-tasking single processor, or a multi-processor device, or even a distributed processing environment - for example, is my brain involved in a reflex reaction? I also wish to avoid discussions about such things as brain and mind. For the sake of this argument, I therefore wish to refer loosely to me, as a sort of "black box being".

The question at stake here then is, do I involve myself in parallel processing situations? The Seymour Papert (1980, pp. 105-112)juggling example (how may I learn to juggle?) presumably indicates that I am. There is also quite a bit of research by the psychologists on this. For example, there are standard experiments investigating what happens if I am given two tasks to do at once. If one of the tasks becomes gradually more demanding, at what point does the other task finally break down? The fact that such experiments are undertaken at all indicate that we have some capability in this area.

An example that to me is fairly familiar, and illustrates matters well is that of learning to drive. One of the early skills learnt is steering; how to keep the car on the road instead of the pavement. This skill is usually acquired fairly rapidly, and all is well - until the need arises to change gear. At this point, full attention is given to the new and demanding problem of operating the clutch and gear lever, and the car starts to wander! As the lamp-post looms ahead, the learner (hopefully!) notices it, raises steering back to a higher level of consciousness, and seeks to sort the tangle out. Eventually, changing gear becomes "second nature", and handling the two processes at once ceases to be a problem. Indeed, the experienced driver may well handle a complex and overlapped sequence of operations with the clutch, gearbox, indicator, brake, accelerator and steering for example, all whilst carrying on a conversation with a passenger. Even more remarkable is the fact that if something goes wrong in one of the processes (eg. the gear has not engaged properly or someone pulls out in front of the car), then this will be raised to a suitably high level of priority in my "scheduling process", and other important processes will continue, albeit at a lower "priority level". Just as with computers, so some other activities - especially talking - may slow, or even stop during such periods of intense concentration.

Are such "parallel processing" scenarios uncommon? I think not; in fact we almost seem to thirst for them. Why do we "doodle" when listening on the telephone? Why do we like to listen to music when working, or doing a jigsaw for example? And many sports test our parallel processing capabilities to the very limit!

It seems to me also that we are rather more complex beings than many of the machines we build. The idea of scheduling processes on a computer system relates to the scheduling that I am doing all the time in the examples above. The idea of interrupts being priority-rated is also something that is inherent in what I do. However, I not only can raise something to a higher priority rating when necessary, I am continually juggling priorities, and sliding things up and down the scale, as in the examples above. Playing a video game is also an excellent example of high level activity and rapid priority adjustment. Whilst priority levels are usually implemented on modern computers, I am not aware of continual priority adjustment being implemented - it would be too complicated!

I therefore conclude that parallel processing is not alien to us, nor to the child learning to ride a bike, nor to the baby learning to walk. It ought to be possible therefore to provide a parallel processing environment that is natural and intuitive. However, all that I have discovered so far on parallel processing is aimed at a very high level, and far from intuitive in my estimation. Presumably, therefore, we have not got it right yet, and this must be worked on.

1.2 I Need Parallel Processing

I suspect that I might meet here the objection that parallel processing is just what I do not need. My first response is to appeal to experience. Again and again, whilst running control workshops, I have seen teachers produce two independent tasks that work well separately. They then ask how they can run them both at once, and receive the reply that they can not. They must design their solution from the outset to contort itself around the need to run and monitor the two tasks at the same time. If current implementations of parallel processing are non-intuitive, the non-parallel solution must be at the far end of counter-intuitive!

An illustrative example is the programming of a buggy with a light to drive round in a figure of eight with its light flashing ("Cops and Robbers"). Assuming that we must solve this with software, it is an introductory-level exercise to make the light flash at an appropriate rate. Similarly, it is not hard to then produce code to make the buggy drive round in a figure of eight. Having successfully coded and debugged the separate tasks, the teacher then wishes to put them together. Note the good modular thinking that we are forever complaining that they don't exhibit! Yet what do we say? It can not be done; stop thinking in this modular fashion and approach it differently.

1.3 Even Simple Solutions Need Parallel Processing

In the quest to research parallel processing, I have been struck by the types of example that are cited as indicating a clear need for parallel processing. Tom Axford postulates a tunnel which for safety reasons can only accomodate a certain number of cars at any one time (Axford, 1989). There is therefore a need for a set of traffic lights at the entrance to the tunnel, a counter at the entrance, and another at the exit. Running the lights constitutes one process, counting the incoming cars another, and counting the outgoing cars another.

However, this is exactly the sort of task that is encountered or devised in a schools context all the time. I do not take this to mean that Tom Axford is wrong; rather, that in schools we are forever trying to make a basically sequential system do what a parallel system would do much better. In this particular example, we would get round it by using a "Counter" primitive, which brings us to the next point.

2 Parallel Processing is Practical

Perhaps parallel processing is desirable, but utopian? The only issue here concerns what we might mean by parallel processing. In fact, "Concurrent Programming" might be a better term to use, refering to the running of computer programs concurrently, rather than the use of processors working in parallel.

This does need brief exploration, since there is considerable theoretical difference between one processor time-slicing between different tasks which appear to the user to be running at the same time, and several processors which really do allow several programs to be active at the same time. This theoretical difference can be hugely important in practice.

An example of this is the use of the "parallel" primitives available in RM Control LOGO. If used for Turtle Graphics, these primitives work provided one is careful to observe standard protocol to avoid corruption of data as the processes switch. If used for control, however, the system becomes unworkable due to the unpredictability of timing which occurs as a result of the switching. Since control is an archetypal real-time system, such unpredictability is disastrous. This is all predicted by standard theory, and in fact, used like this, the system feels like the sort of system that accompanies the further reaches of computer science, rather than the primary classroom.

In practice, many sequential systems available to schools have what I would call "parallel fixes" in order to cope. One example is the "Counter" set of primitives mentioned previously. These allow the setting up of a counter to count pulses on digital input lines, and which feel rather like those used to access the real-time clock. Thus the counter can be set to zero, and it will then appear to work away in the background, counting pulses as they appear on a given input line. This total may then be accessed in the foreground when required. It is this set of primitives which make Tom Axford's "Tunnel" example above relatively easy to implement on a school system.

In terms of practicality, it is also important to note the speed with which computer technology is advancing. It is now the accepted norm that cheap personal computers work in a windowing, multi-tasking environment, with full concurrent programming coming in rapidly. The machines are already so fast that the question of whether we will see transputers bringing true multi-processing into common use is almost irrelevant. They appear to the user to multi-process anyway. It is also interesting to note that multi-tasking was not only possible on the old British "BBC" computer - it was used. There was a control program called "Controller" which provided an important clock facility, and its screen display worked in multi-tasking mode, giving a real-time display of input status, whatever else the program was or was not doing.

2.1 Do I Have to Do It by Software?

Another issue tangled up in this is what might be called software/hardware equivalence. In the "Cops and Robbers" example above, a simpler solution by far is to use a light that flashes, rather than one which must be made to flash by software. In this case, one simply switches on the flashing light and forgets about it; all attention can be on driving the car around. In virtually all introductory control scenarios, such alternative solutions can be posed, making the software solution trivial, or indeed, making the use of a computer altogether unnecessary. Whilst such considerations, and the methodology for deciding what kind of approach to use, are important, they are not relevant to our current problem. The point is that software solutions rapidly lead us to a need for parallel processing. If the software experience is valid and important, then the parallel processing bit is important too.

2.2 What Sort of Parallel Processing Do I Need?

I believe that it is very easy to follow the conventional lines in the realm of Concurrent Programming and then find that we can not see the wood for the trees. It is important to go back to first principles and decide what the constraints really are. If I consider that RM Control LOGO is not a suitable starting point for parallel processing for a child, then what is?

An important issue to refer to is that of "Visibility" (Du Boulay; Argles, 1990, p. 79; and Argles, 1991, p. 263). A process can be too fast for its own good, as far as the learner is concerned; I may need to slow it down. In fact, the amount of processing time needed for most introductory control applications is minimal. By far the greatest amount of time is spent waiting either for a time lapse, or for a change in status of the inputs. This is an ideal scenario for a time-slicing solution. Why bother with adding the complication of extra processors if the one we have got is standing idle for 99.9% of the time?

Two mechanisms (both commonly available for the last ten years on home computers) are required to take advantage of this. The first is the use of a real-time clock to implement time delays, rather than wasteful program delay-loops. The second is the implementation of interrupts, so that "events", such as the completion of a time delay on the clock, or a change of status of the input lines, can be implemented. These mechanisms are those used to implement the "Counter" primitive, for example, and it should be relatively simple to use the same technique to ensure that the timing of concurrently programmed control procedures is not significantly affected, even when run on a single processor.

3 Communication Between Processes

Another huge area of complication in concurrent programming is that of interaction between concurrent processes. If two processes need to interact, how should this be achieved? By "Rendezvous"? By sending "Messages"? I would argue that this is an unnecessary "if", certainly in the initial stages. The examples that I have come across in practice do not require communication between processes, only a supervisory scheduler. Thus I envisage the programming and debugging of individual tasks to be done at "Top Level" in LOGO terms, and then left in program memory for invocation when required. Invocation should then be made by a "Scheduler Program" which can start it or stop it as required. This thinking fits best with the "Windows" approach as used in SMALLTALK 80. Tasks could then be written in the highest level window, but "sent to a lower level window" when completed. This leaves the highest level window available for writing the Scheduler when required. Alternatively, one could start with the Scheduler, and open up lower-order windows as necessary for the constituent tasks.

If the need for communication between processes should arise, the object-oriented philosophy of sending messages should prove to be appropriate, without introducing unnecessary complication. At any rate, such a philosophy fits with the windows approach, and there is an exemplar in SMALLTALK.

Easy entry into a system for a novice is most important, especially if our approach is to be accessible by a child. Another advantage of the approach outlined above is that parallel processing need not intrude at all in the early stages; a simple sequential process can be defined, tested and debugged, all at top level with no idea of scheduling going on at all. Then, when a parallel approach is required, it can all be parcelled up in a window environment, and scheduling can be introduced without making a big fuss about it, or immediately introducing technical jargon.

4 How Might I Access Such a System?

I have stated a personal view that conventional parallel programming environments are unfriendly and frequently non-intuitive. I have also stated that I feel a windows approach makes things easier to deal with. This brings us into the second of the major issues raised at the start of the chapter; what kind of program interface is most appropriate for this envisaged system?

4.1 There Is More to Graphical Interfaces than Meets the Eye

An early exploration that I made concerned the Concept Keyboard, which is a paper-sized (A3 or A4) redefinable keyboard. A problem with a command-driven progamming environment like LOGO is that a young child has problems with spelling and syntax. Seymour Papert's solution was to introduce a "Button Box"; mine, because it was there, was to use the redefinable function keys on the computer keyboard to provide frequently-used commands, complete with separating space, with a single key-press.

Various isues are involved here, such as the distinction between just providing single commands with a keypress, and Seymour's approach which was rather more subtle (defining colours, etc.). Nevertheless, the function-key approach was effective, and has remained in place from the start. Other software writers have also copied this idea fairly closely, which at least should indicate that it is a reasonable solution within the current constraints of practice.

Later, the opportunity arose to encourage the development of a set of overlays for the concept keyboard which allowed considerable flexibility, range, and graphical approaches to be used. The design of the overlays was not undertaken by me, and clearly is important. However, the effect that surprised me most when I trialled the system was not overly affected by this.

The function keys are clearly a part of the computer's keyboard, and when pressing them, the children expected to look at the screen first to observe the effect. The Concept Keyboard is separate from the computer, however, and looks and feels much more like the keyboards available on programmable toys, where no visual feedback (in programming terms) is available. The children would press things on the Concept Keyboard, therefore, and look directly at the control model for an effect. The program line on the screen was not therefore looked at, and any messages back from the system about errors of logic (or syntax) were missed. As a consequence, the introduction of the Concept Keyboard actually hindered younger children's development with the system, rather than helping them.

4.2 What About Icons?

The issue of what sort of icon to use, initially brushed aside in this consideration, is actually most important. Icons can be powerful carriers of meaning, and frequently they imply an underlying metaphor. If they are used, therefore, great care must be used to ensure consistency and appropriateness, else they become confusing. In my opinion, the British Archimedes computer has some excellent examples of misleading icons which totally mystify the new user, e.g. the RAM disc icon which bears no relation to the physical disc drive icon, although their function is similar.

I can also see that iconic approaches to programming, or Visual Programming, could be immensely helpful aids to understanding complex programs. At some stage, it might be helpful to consider how such thinking could helpfully mesh with the windows thinking outlined above. Certainly the statechart approach as outlined by Harel (1987), perhaps with "layers" of complexity implemented within the icon environment, could be a useful starting point. However, I wish to start with the right approach for the novice or the child, and complexity of program will not be an issue at this level. I therefore wish to leave this ground "lying fallow" for the time being.

4.3 What Sort of Windows?

At the moment, I simply envisage the windows environment being a convenient way of "parcelling up" what I would like to call "Control Tasks". Thus LOGO might start up in a window, say. My task of making a light flash is then implemented by opening up a sub-window in which I program a flash procedure, perhaps. This task might be tested by returning to the "parent" window, and using a super-procedure to tell the sub-task to start.

In order to make my buggy travel round in a figure of eight, I now open up a second sub-window, in which I program the figure of eight procedures. This can also be tested by returning to the parent window.

If I now wish to run both tasks together, I simply alter the program in the parent window to tell both tasks to start at once. This has effectively become my Scheduler. Note that in this simple example, there is no need for the tasks to communicate with eachother, nor even for them to communicate with the scheduler. There is a need, however, for the scheduler to be able to interrupt a task that it may have initiated, and which may not have terminated.

It seems to me that the implementation of such a windows-based approach would facilitate the modular thinking that we are keen to encourage as good practice. It would also seem to be a way of introducing concurrent programming in a reliable and easy to understand way. But this needs proving in practice.

It is also worth considering that the "package" set of primitives fit in with this sort of thinking. However, I think that the window environment makes it much easier to cope with, extends the thinking, and can force such structures from the beginning, rather than making it "available if required".

5 What About Communication Between Tasks and the Scheduler?

The event-driven approach to control seems to me to be a helpful way forward for implementing communication between parallel tasks in the control environment. It also seems to relate well to the way I operate as a human being.

If we think back to the driving example at the beginning of this paper, the way in which I operate as a human being seems to me to feel like an interrupt scenario. Steering becomes a lower level task while I concentrate on changing gear in the early stages. However, as the car begins to wander across the road, hopefully my vision registering a lamp-post looming ahead, or failing that the yell of my instructor in my ear, provide an "interrupt" that tells me to get steering back up to being a "top level task".

Similarly, the fact that we are forever being exhorted to be "pro-active" rather than "reactive" suggests to me that we are naturally reactive beings, responding to stimuli or "interrupts".

As a first step, I therefore envisage communication being controlled by the scheduler, but being on an interrupt basis. A constituent task may therefore sense something important, and send a message back to the scheduler. The scheduler may not be listening, or it may choose to ignore it, or it may listen and act; but that is the scheduler's responsibility. All the task has to do is send the message.

5.1 What Might this Mean in Practice?

As I have sought to write this down, my thinking has changed. I have taken the generalised case of tasks that sense and inform the scheduler of relevant changes in incoming information. I have tried to think of examples to illustrate this, and every one has reduced to a simpler scenario. The tasks reduce to ones that do not need to communicate, and the sensing reduces to a task that says, in effect, "When this condition is detected, tell the scheduler" - in other words, what might be called a "When demon". This is most important, because it has been noted elsewhere that there is a tendency to provide complicated solutions for simple problems (Argles, 1990).

5.2 Responding to Inputs

If we are going to consider WHEN demons, it is also important to consider how it might be implemented. At first sight, DOUNTIL appears to offer much in this context. However, all is not as it seems. The intuitive concept would appear to be that the specified task is run continuously, but is stopped immediately the specified condition occurs. Thus:

DOUNTIL [SWITCHON 1 PAUSE 1 SWITCHOFF 1 PAUSE 1][INPUTON? 2]

The first list tells, say a light, on line 1 to flash on and off with a mark/space period of 1 second. Perhaps the second list specifies that this will be halted as soon as the sensor on line 2 switches on? Not so in BBC LOGO. What actually happens is that the first list is run, then the test is made, then the first list is run again and so on. In practice, the command list will always complete a cycle, and the change in state of the input sensor must be quite (unacceptably) long if it is to register. DOUNTIL is therefore hardly the concurrent programming primitive that it first appeared to be.

5.3 The Problem with Interrupts

What we are really looking for is the ability to initiate a task, and then to interrupt it in "mid-flight". This, however, introduces new problems. After an interruption, what happens next? Do we pick up from where we left off, or do we give up and do something different (or stop altogether)? The conventional approach to computer interrupts is for the current task to continue where it left off before the interrupt occured, but with a facility to pass a message back, together with a possible "read me now" flag set. This again seems a reasonable approach.
   The default can therefore be a WHEN primitive that says WHEN this condition is detected, do this, and then go on as before. If we propose a STARTTO primitive to initiate a predefined task, our scheduler program could perhaps look like this:

        STARTTO [FlashLight]
        STARTTO [DriveRound]
        WHEN ObstacleDetected? [AvoidIt]

This would start the procedure to flash the light, then start the procedure(s) to drive the buggy round in a figure of eight, say. The WHEN line then sets up, in effect, an interrupt routine that says "When the bumper sensors detect an obstacle via the ObstacleDetected? routine, take avoiding action by running the AvoidIt routine."
   A number of issues are raised here. Firstly, wouldn't it be better to change the WHEN primitive to WHENEVER, on the basis that, in the general case, we will want it to remain active? I think this is probably true, but again, it needs testing out.

5.4 Unplugging Interrupts

Having set up WHEN, or WHENEVER, how do we unplug it again when we no longer are interested in looking for a certain condition? The computer parallel to WHENEVER is a command that changes an "Interrupt Enable Flag". It is ObstacleDetected? that is sending the message; WHENEVER is simply enabling the message to get through. We could therefore have an IGNORE command, for example, that could be used like this:

IGNORE [ObstacleDetected?]

This would simply mean that the messages from ObstacleDetected? are now to be ignored. IGNORE has been used in Archimedes LOGO to allow the invocation of an operation, whilst ignoring the output from it (eg. remove the next character from the input buffer, but I'm not interested in what it is). This needn't be a problem, especially as the two are related, but it leaves a subtle distinction between:

IGNORE [ObstacleDetected?]

- ie. undo a WHENEVER, and IGNORE ObstacleDetected?

- ie. run ObstacleDetected? once and ignore the output.

The implication of the above is that, just as STARTTO initiates a concurrent task, so WHENEVER also initiates a concurrent sensing task - in this case, called ObjectDetected? Initially, one would expect this to reduce to a single operation, so:

WHENEVER INPUTON? 3 [AvoidIt]

However, it is important to realise that this masks the fact that continually sensing input line 3 has become a concurrent task.
    It is also important that the sensing task should be able to implement a software "Escape" function if required, together with a THROW and CATCH scenario.

6 Implementation

I suppose it is a fact of life that there will often be a gap between what one wishes to develop and what is feasible in practice at any point in time. Since putting these thoughts on paper, work has begun on a program to implement some of these ideas. There are many things one would like to implement, but the constraint of resources forces a more restricted target.
    The following implementation is seeking to provide a product that will enable the main thesis to be tested; that with the right approach, seven-year-olds can and should write concurrent programs for control.

6.1 The Design

The control hardware consists of an Archimedes ARM3 computer connected to a serial interface with eight input lines and eight output lines. Communication is at 9600 baud. The various elements of the hardware have been developed over a period of time and have been proved to be reliable.

6.2 The Software

The software is of an object-orientated design working in a windows environment. On running the application from disc, two programs install on the icon bar at the base of the screen. One appears on the left of the bar and is effectively a "device driver" for the serial interface. It receives information from other applications for output to the interface, and holds information from the interface ready for collection by other applications. It also configures the computer to match the interface.

The second module installs on the right of the icon bar and provides windows in which the child may write control programs.

A modular approach towards development is being adopted, so at the point of writing this paragraph, the icon provides windows, and each window has a text editor to enable programs to be written. The interpretter still has to be written; hopefully, it will be ready in time for the delivery of this lecture.

The basic programming syntax is a LOGO variant. For the purposes of testing the thesis proposed, it could be a "toy LOGO", although it would be nice if it could be developed.

6.3 Using the Software

When a window is initiated from the icon on the icon bar, a text file may be dragged to it in order to load a pre-defined program. Alternatively, the text may be typed in within the desktop and edited in a conventional way.

If the menu button is pressed on the mouse, a menu provides opportunities to save the program as text, to run a named procedure, to run from the top of text, pause execution, or continue.

As the program runs, various program primitives such as SWITCHON and SWITCHOFF communicate with the interface via a "hot link" to the interface program mentioned before.

6.4 Communicating Tasks

The contention is that concurrent programming need not demand the use of conventional "rendezvous" structures, etc. in the initial stages. However, there is a need to provide synchronisation structures ready for use when they become necessary. In order to allow windows to talk to each other, it is envisaged that additional primitives will be required. These might be as follows:

TELL <window> [message]	sends [message] to <window>.
HEARD	waits for, & returns a TELL message.
HEARD?	returns TRUE if a TELL message is waiting.

Any synchronising action is then the responsibility of the window receiving such a message. Further primitives that might allow such synchronisation to occur could be as follows:

BEGIN [procedure]	starts up [procedure] as a concurrent process.
STOP [procedure]	suspends execution of [procedure].
RESTART [procedure]	restarts [procedure] from where it left off.

7 Conclusion

In this paper, I have argued that control scenarios are inherently parallel in nature, that people are parallel-processing by nature, and that there is a mis-match between the single-tasking programming languages we provide children with and the real world.

Various aspects of design have been explored, and a practical design brief produced for a concurrent programming environment for children. This design is being implemented.

Following this, it is planned to trial the prototype and to refine the basic design, so that the basic thesis may be tested. It would be useful to produce a version of the software for the IBM PC; however, the programming expertise and necessary resources to achieve this are not yet available.

Acknowledgements

Thanks are due to Dave Bruner, my research supervisor; to Maurice Meredith, Derek Bunyard and Ron Allen for their helpful prodding; to Helen, my long-suffering wife; to Haven, Blake Electro Production Ltd and King Alfred's College for their practical support; and especially to Ting Quei who has turned windows programming from a pipe-dream into a reality.

References

Axford, T.: Concurrent Programming, Wiley, 1989.

Argles, D.: A Communicating Logo. In: Schuyten G., Valcke M. (eds.): Teaching and Learning in Logo-Based Environments, IOS, 1990.

Argles, D.: Logo and Control. In: Calabrese E.(ed.): Third European Logo Conference. Proceedings, Parma 1991. Du Boulay et al: The Black Box inside the Glass Box. In: International Journal of Man-Machine Studies, 14.

Du Boulay et al.: The Black box inside the glass box, in International Journal of Man-Machine Studies, 14.

Harel, D.: Algorithmics - The spirit of computing, pp. 294-300, Addison-Wesley, 1987.

Papert, S.: Mindstorms, pp. 105-112, Harvester Press, 1980.

Learning Mode in the Exploration of Parallelism in Pedagogical Robotics

Jean-Baptiste La Palme[1] and Maurice Bélanger[2]

[1] Département de Mathématiques et d'Informatique, Université du Québec à Montréal, C.P. 8888 Succ"A", Montréal, Canada, H3C 3P8

[2] Centre interdisciplinaire de recherche sur l'apprentissage et le développement en éducation, Univ. du Québec à Montréal, C.P. 8888 Succ"A", Montréal, Canada, H3C 3P8

Abstract. In learning mode a student can, at the same time, control and program a robotic device.

This mode offers the student some advantages: the immediate feed-back of commands entering the program, the ease and speed of programming because timing between commands is implicit, and the possibility to run a program while still maintaining immediate control of the device.

These characteristics of the learning mode help the student explore parallel programming and parallel control: the interweaving and nesting of processes, the way processes can control the termination of other processes, and the relations between sequential and parallel processes.

Keywords. Pedagogical robotics, Robotics, Learning mode, Teaching mode, Parallelism, Education, Programming, Concurrent programming.

1 Learning mode

Learning mode, sometimes called the teaching mode, is used in the programming of robotic devices where it is faster, simpler or cheaper than ordinary programming techniques.

Different kinds of learning modes are used in industrial robotics, from mimicking, as in a paint shop, to showing to the computer some important points of positioning, as in a welding shop. (Safford 82, Blume 86)

The "structured learning mode" we developed for 5 to 15 year old students relates to both of these examples.

In some situations, the elapsed time between the pressing of two keys will determine the timing between the execution of a command and the execution of the next one.

In other situations, the timing between the execution of commands will depend on events: the change in state of a sensor, or the termination of one or several processes.

The word "structured" is used to underscore the fact that we decided to favor a hierarchical organization of learned processes, by permitting nested control commands only via the definition of processes.

2 Implementation

We chose to implement this learning mode by transforming the function of the screen and the keyboard. [see fig. 1]

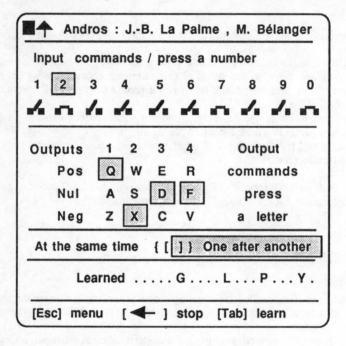

Fig. 1. Control panel

Keys of the keyboard are given different functions: some keys activate and deactivate outputs, some keys start waiting loops for an event, other keys activate learned processes and others change the mode of operation.

The screen provides visual feedback of the system: state of outputs and inputs, state of processes and waiting loops, modes of operation.

There are four non-exclusive modes of operation: execution mode, learning mode, sequential mode, and parallel mode.

This system has been used with robotic construction sets for 12 to 15 year old students, and with robotic puppets for 5 year-old students. The first year, we used hand puppets and the second year, we used wire-puppets [see fig. 2].

Fig. 2. Hand-puppets and wire-puppets

3 Elements of structured learning mode and some notations

In this structured learning mode, we use the following notions:
1° "output command" activate or deactivate robotic outputs
2° "input commands" are waiting loops ending when an event occurs
3° "processes" are sequences of the following optional elements: input commands, output commands and processes.

4° "control commands" modify the order of execution of elements of a process (sequentially or simultaneously)

The following notations are used for output and input commands:
    1° Q and W are output commands activating outputs 1 and 2 respectively (the choice of letters depends on the place of letters on the QWERTY keyboard).
    2° A and S are output commands deactivating outputs 1 and 2 respectively (the choice of letters depends on the place of letters on the QWERTY keyboard).
    3° 1,2,3,4 are input commands ending when switches 1,2,3,4 respectively change state.
    4° +1,+2,+3,+4 are input commands ending when switches 1,2,3,4 respectively are on.
    5° -1,-2,-3,-4 are input commands ending when switches 1,2,3,4 respectively are off.

The following notations will be used for sequential control commands:
    1° < a b c > is a process in which the elements a b c are executed sequentially.
    2° ≤ a b c ≥ is a process in which the elements a b c are executed sequentially endlessly (infinite loop)
In learning sequential mode, the elapsed time between the pressing of an output command and the next command is recorded, and determines the timing between these commands at execution time.

The following notations are used for parallel control commands:
    1° [p q r] is a parallel process in which the processes or input commands p q r are executed simultaneously. This process ends when one of its elements ends; the other elements are then interrupted.
    2° { p q r } is a parallel process in which the processes p q r are executed simultaneously. This process ends when all of its elements have ended.
When defined, new processes are given a single-letter name that becomes another single-keystroke command: the notation p : <... elements...> or p : ≤... elements...≥ or p : [... elements...] gives the name "p" to the process.

4 An example

A typical example of first experimentation with this system would be an alarm system composed as follows: a magnetic switch triggering the alarm for a limited time, a switch that may disable the system, and a flashing light indicating that the alarm is activated.
    The flashing of the light is a process that can be defined by the following sequence L (if the light is connected to output no 2):
        L: ≤ W S ≥
The alarm is a process that can be defined by the following sequence M (if the bell

is connected to output no 1, and the magnetic switch is connected to input 2)

    M :   ≤ 2 Q A ≥

Then the alarm system G could be defined by the following sequence (if the switch to disable this system is connected to input no 3):

        G :   < [L M 3] P >

where P is an output command that deactivates all outputs. P is necessary, because when L and M are interrupted, we do not know in what state the outputs 1 and 2 are left.

It should be noted that during the pressing of the keys for these sequences of commands, the corresponding commands are recorded and executed and echoed on the screen, giving the user a robotic feed-back and a visual feed-back of his actions.

5 Interweaving and nesting of parallel processes

In more complex situations, the visual feed-back on the screen of active processes greatly helps understand the dynamics of parallel processes.

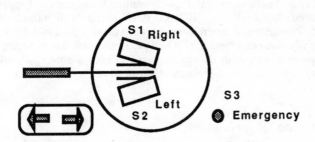

Fig. 3. Flashing light indicator system

For the sake of clarity let us take an example that everyone knows: the flashing light indicator system of a car. The right light flashing for right turns, the left light flashing for left turns, and both flashing for emergency situations. We will construct a simulation of this system (see Fig. 3).

The infinite loop process that makes the right light flash could be defined by:

      L :   ≤ Q A ≥

The infinite loop process that makes the left light flash could be defined by:

      M :   ≤ W S ≥

We can note that in both these processes, there is no provision for the beginning nor the ending of the flashing. This is coherent with the fact that these processes could be triggered or interrupted by many different processes.

One process that triggers the right light flashing (the indicator lever up, on input no 1) could be defined this way:

K: ≤ +1 [L -1] P ≥

One process that triggers the left light flashing (the indicator lever down, on input no 2) could be defined this way:

N: ≤ +2 [M -2] P ≥

And the process that triggers both lights flashing (the emergency button down, on input no 3) could be defined this way:

U: ≤ +3 [L M -3] P ≥

Since the left and right flashing are incompatible with emergency flashing, we could define the direction flashing system this way:

O: ≤ -3 [K N +3] P ≥

Then the overall system (switch 4 for disabling) could be defined by:

G: < [O U 4] P >

For many users, it is quite unexpected that, when the right indicator is activated, all processes are active except M, when the left indicator is activated, all processes are active except L and when the emergency system is on, all processes are active except K and N.

The screen visual feedback (active processes shown in inverse characters) and some partial tests help the user to realize, that:

- a process could be active, even if it is only waiting for an event to happen.
- when nested chains (GOKL) (GONM) (GUL) (GUM) of infinite process are activated simultaneously, interrelations between them depend mainly on events.

More generally, these experiences help students realize that it is more useful to see a parallel system as a living body than to see it as sequence of operations, and thus it is more useful to concentrate on its processes than on its states.

6 Termination of processes

As we have seen, the termination of processes needs not to be determined when these processes are defined, but if we wish, only when they are combined with other processes. We have seen how events can be used to modulate their termination conditions. We will now construct an example where the duration of a process determines the termination of an other process.

Again we shall take an example from an automobile. We shall construct a simulation of a seat-belt alarm: when the contact key is turned, the system verifies if the driver and front passenger have buckled-up.

Since a passenger may or may not be present, there is a switch that verifies the presence of a passenger. If one person does not have the belt fastened, then for a certain time a warning light glows and a intermittent sound alarm is activated .

The intermittent sound alarm may be defined as an infinite process (the alarm is connected to output no 1):

M :   ≤ Q A ≥

The glowing of the light may be defined as a process whose duration will determine the maximum time during which the sound alarm will be activated (the light is connected to output no 2):

L :   < W S >

Now, we can define a process whose duration will depend on how long it takes to fasten the seat belts. Suppose switch 1 verifies the belt of the driver, switch 2 verifies the belt of the passenger and switch 3 verifies the presence of a passenger.

The process that verifies the passenger belt could be defined by:

K :   < [-3 +2] >

this process will end if there is no passenger or as soon as the passenger's seat belt is fastened.

The process that verifies if all the seat belts are fastened is then:

N :   < { +1 K } >

this process ends when the driver's seat belt is fastened *and* when K is finished.

Then the seat belt alarm system would be:

G :   < [M L N] P >

We can see that the duration of the sound alarm would be the same as the duration of the shorter of the two processes L and N.

Most readers at this point would have invented another solution using a programming logical expression such as (1 and (2 or not 3)). This would, of course, imply that there is somewhere an "if" control structure included in an infinite loop built by the user.

Since this "structured learning mode" is built around input commands that wait for an event instead of testing for a condition, the "when" is favoured to the "if", and "dynamic" thinking is favoured to "static" thinking, directing the student to explore the time dimension.

7 Sequential vs parallel processes

In previous articles we have presented some examples of fairly complex tasks that can be resolved more easily with expressions involving parallel and sequential processes (La Palme and Belanger 91).

We do not know to what extent students at this level may gain insight on the dialectic between parallel and sequential processes.

We do know that parallelism per se does not represent a greater difficulty in solving parallel problems than sequentiallity in solving sequential problems.

Furthermore, we have seen numerous examples where students did not have difficulties in coming up with expressions equivalent to:

{ < p q > < m n > } t

Less often we have seen expressions equivalent to:

< { p q } { m n } > t

However, students sometimes have to resort to common sense when confronted with problems mixing the two modes with infinite processes. For example, in solving "train" problems some students came to realize that in a sequence equivalent to

{ ≤ p q ≥ ≤ m n ≥ } t

the process t would never be executed. But, they would not think of writing a sequence equivalent to

< ≤ p q ≥ ≤ m n ≥ > t

Since it was so easy and fast to combine parallel and sequential processes in this system, students did not hesitate to explore different formulations of the same process. *They came to see the system as an "editor of movement", which is a nice way to regard the activity of programming robotic devices.*

8 Pedagogical example

We present the students with the following problem: a block is placed on a train by a push rod conveyor, the train goes to a predetermined platform, unloads and comes back, the conveyor is recharged with an other block and the process is repeated endlessly.

Sub-processes can be defined as follow: [see fig. 4] L charges the block on the train. N brings the train to the unloading platform and unloads. M brings the train back. O brings the conveyor's push rod back to permit the loading of another block.

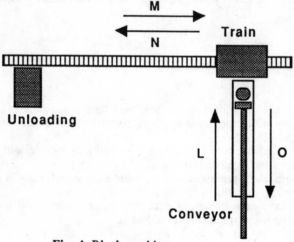

Fig. 4. Block problem

The first solution of the students is generally sequential and could be described as:

G :   ≤ L N M O ≥

After a while, students realise that the process is time consuming and that some movements could be executed simultaneously.

Generally, the first movements that are picked to be executed simultaneously are N and O. So the solution is then changed as follow:

H :   { N O }
G :   ≤ L H M ≥

As if the situation was symmetric, some students tried to do the same thing with L and M, leading to the following solution:

H :   { N O }
K :   { L M }
G1 :   ≤ H K ≥

but then, since L was faster then M, the block fell to the ground, because the train was not yet in the proper position for receiving the block. At this point, students found a mechanical solution by replacing gears on the conveyor in order to reduce its speed, allowing the train more time to come into position.

At the end of the class-period, the gears were left in the slower configuration. So, when students of the following group were presented the same problem, they invented yet another solution, seemingly taking into account the new configuration:

H :   < N M >
K :   < O L >
G2 :   ≤ { H K } ≥

The two solutions G1 and G2 are consequences of two very different mental representations of the problem, and are written in two very different forms: one being a parallel process composed of two sequential processes, while the other is a sequential process composed of two parallel processes. However, the processes G1 and G2 give almost the same results concerning the blocks, the only difference being the waiting time of the train: in G1, the train waits at both the unloading position and at the loading position, while in G2, the train waits only at the loading position.

There is a big operational difference between G and G1,G2: the starting position. In G, the push rod must be in rear position, while in G1 and G2, the push rod must be in front position.

We do not know how these last considerations affect the conceptualization of parallel processes or sub-processes.

Acknowledgment

The research for this article was funded in part by the CRSH of Canada.

References

Banâtre, J.-P. (1991). La programmation parallèle, outils, méthodes et éléments de mise en oeuvre. Editions Eyrolles, Paris.

Blagoev, L. Nyagolova, L. (1989). Two levels of parallelism or a new approach for control system design. 1989 FORML Conference Proceedings, F.I.G. San Jose, CA.

Blume, C., Jakob, W. (1986). Programming languages for industrial robots. Springer-Verlag.

Krishnamurthy, E.V. (1989). Parallel processing, principles and practice. Addison-Wesley.

La Palme, J.-B. (1990). Un théâtre robotisé de marionnettes, construit et programmé par les élèves à l'aide du langage Androïde. Bulletin du Réseau Informatique et Education. APO Québec, Ecole Internationle de Bordeaux, Université Mohamed V. Dec.

La Palme, J.-B. Bélanger, M. (1991). The implication of parallelism in pedagogical robotics 3rd International Conference On Pedagogical Robotics, Mexico.

Safford, Edward L. Jr. (1982). Handbook of advanced robotics. Tab Books, PA.

Terry, C. Thomas, P. (1988). Teaching and learning with robots. Croom Helm.

ActNet - A Heterogeneous Network of Actors for Learning of Parallelism, Communication, and Synchronization

Philippe Darche (MASI laboratory) and Gérard Nowak (LITP laboratory)

Université Pierre et Marie Curie, Institut Blaise Pascal, couloir 65-66 - pièce 101, 4, place Jussieu, F75252 Paris Cedex 05, France
E-mail : darche@masi.ibp.fr and nowak@litp.ibp.fr

Abstract. We are proposing a tool designed to help apprenticeship of parallelism and communication, for different kinds of audiences such as schools, colleges and training centers for adults. The system is based on the "micro-world" concept defined by Seymour Papert. Graphical simulation appears insufficient for achieving this purpose so we prefer an interactive environment: a network of real hardware and software actors. We use some 'pedagogical robotic' materials, in particular static and mobile actors. Each actor is equipped with a board designed with a transputer, running a Lisp interpreter. We present our hardware and software choices respectively for this project which is still in progress. Afterwards two contexts using this environment are shown.

Keywords. Actor, Control technology, Microworld, Network, Parallelism, Problem solving, Programming, Pedagogical robotics, Transputer.

1 Introduction

With architectural advances of computing systems, parallelism is becoming a normal control structure for actual programming applications. Nowadays, the human mind is mainly moulded from childhood to understand purely sequential processes, therefore the comprehension of concepts bound to parallelism and mastery of associated mechanisms are not easy to achieve.

Inside the context of apprenticeship of these concepts by pupils, students or adults, we propose a tool, using little pedagogical robots, which enhances co-ordination by synchronisation, communication, or concurrence, competition and mutual exclusion between parallel processes. This study led us :
- firstly, to equip these robots with resources (processor, memory, communication interface) and to create a network of hardware communicating actors,
- secondly, to give a software character to these actors, by providing them with a Lisp interpreter, with the aim of opening the field of potentially parallel applications runnable on the net to the favoured domains of this language: objects, actors, agents and expert systems.

The global product is intended to be used for apprenticeship of the fundamental mechanisms bound to parallelism in an industrial context as well as in a scholastic system.

We have designed and are developing an Inmos transputer based board, supporting a "home made" Lisp system (Méta-Vlisp), with the aim of equipping each kind of the chosen robots (sedentary or mobile). Becoming autonomous in this way, robots are, thus, hardware and software communicating actors.

Within this framework, we are now exploring the possibility of implementing Méta-Vlisp with ADA onto transputer, and integrating this Lisp system in an ADA environment of packages and tasks. We therefore hope to make Lisp and ADA systems mutually beneficial.

2 Hardware Architecture

The ultimate step in the conception chain of a distributed application is its 'in situ' execution. In a pedagogical context, the software simulation requires a large capacity of abstraction from students; so, we preferred experimentation with hardware devices, by using a specific micro-world [Papert, 1980]. Indeed, micro-world concepts have been revealed to be tools well suited to apprenticeship, favouring interaction by the observation of precise phenomena. A hardware micro-world, according to our proposal, is extended to a set of pedagogical robots, driven by a personal computer, and communicating with each other; we associate this set with another set of passive objects (for example: cubes, places, communication ways, varied obstacles). The user interacts with the environment by sending commands to an interpreter of a language.

2.1 Robotic Actors

The use of pedagogical robots is justified by the fact that they are robust, with a good quality/price ratio, and that their performances, in spite of reduction factor, are real. We classify the pedagogical robots in three categories:
- the robots of 'industrial' type, fully functional, physically reduced reproductions of existing industrial robots,
- the robots of 'educative toys' type (LEGO®, Fischer-Technik®, etc.), only reproducing a functional part of reality, and
- the robots issued from research (T3 mobile robot from Jeulin).

The latter often exist as prototype models (we quote for example the LOGO "turtle"-like mobile for psycho-pedagogy). Our choice leads us to select two kinds of robots: first, a sedentary robot with a six degrees arm of liberty, named 'Youpi' [J.D. Productique 86] and already chosen for colleges and high schools; and second, a mobile prototype robot of the LOGO "turtle"-like type, that we designed and which is currently under construction. The mobility of a robot implies energetic autonomy and telecommunication capacity, hence the following architecture.

The mobile robot is composed of objects' groups:
- the propulsion group composed of two step-by-step motor type objects,
- the printing group composed of a pencil type object,
- the sensor group composed of six collision sensor type objects (proprioceptive sensor),

The power supply group composed of:
- a battery's charger type object,
- a battery type object,
- a DC/DC converter type object,
- an external power supply object,
- the telecommunication group composed of a V.H.F. modem type object,
- and the data processing group composed of a generic command board type object.

The use of others types of robots (crane runway, endless belt) is considered for being able to simulate, afterwards, a complete flexible manufacturing system.

The notion of actor is prompted by the social organization's model of the communicating experts [Hewitt, 1977]. It can be considered as a co-operative and distributed computation model with an 'objects' approach [Agha, 1986]. In a micro-world context, the robots will be seen as real hardware actors [Darche, 1991]. The hardware size of the project needs to extend the standard actor's Agha model. We specify and synthesize the notion of actor in our project by means of a grammar.

An actor can be defined by the following rules :
    actor --> software_package [hardware_package]

This point of view is integrating the standard notion of actor :
    actor --> standard_actor | hardware_actor
    standard_actor --> software_package
    hardware_actor --> software_package   hardware_package

2.2 The Network of Actors

As the control structure in an actors's model is message passing, the hardware actors are organized in a local network named "ActNet" for "Network of Actors"[1]. It is split into three sub-networks (see Fig. 1.1 and 1.2):
- the users' sub-network (U.S.N.), dedicated to the development and control of users' applications and,
- the executive sub-networks of moving actors (M.A.S.N.)
- sedentary actors (S.A.S.N.).

The network is heterogeneous by its topology (programmable: M.A.S.N, ring or bus), but also by its kind of processors equipping every node.

The USN is a local commercial network with a variable topology defined by computers supporting the developments of the users. In this study, it is reduced to a Personal Computer equipped with an interface board for SASN.

---

[1] The reader can find a glossary referring to some abreviations at the end of this paper.

The SASN is a double ring, with opposite directions, to increase reliability and insure a high and warranted speed for the flow of data.

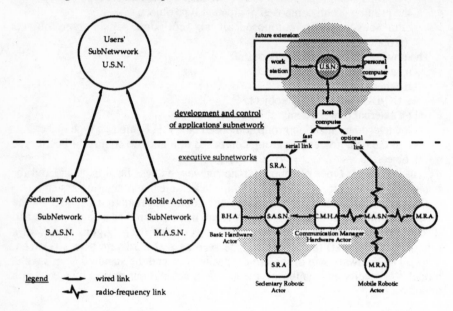

Fig. 1.1. ActNet Topology   **Fig. 1.2.** Details

The MASN is a radio frequency network (V.H.F.) to guarantee the autonomy of mobiles with communications [Darche 91]. We want an asynchronous full duplex communication. The modulation speed is the one of a serial RS232C type link equipping the control board, and this one is setting from 1200 to 19200 bauds. A channel coding is needed to adapt the information. To achieve a duplex communication, it needs to spot the information coming from several interlocutors. An analogue modulation will be used. The frequencies' space will be cut in identical bands called radio-frequency channels and two channels will be allocated to each actor of the MASN, which will be therefore characterized by two frequency addresses, one for the emission and one for the reception. For spectrum occupation and reliability, a frequency type (F.M.) discrete angular modulation, the binary minimum frequency shift keying modulation (B.M.F.S.K. modulation) has been chosen, because it is optimizing the channel bandwidth necessary for the transmission of a digital signal, without an elaborated source coding. The use of digital technologies (Phase Lock Loop particularly) allows us to compute sending and receiving frequencies of each mobile. The use of an antenna's duplexer ensures full duplex communications, but because of duplex spacing between the emission and reception frequencies, two distinct frequency bands B1 and B2 must be defined. For legislation reason, we have chosen :
- B1 = [40 , 45] Mhz,
- B2 = [70 , 75] Mhz.

The wanted maximum modulation speed and the employed type of modulation are defining the channel width (50 Khz).

So these technical choices enable us to obtain the best fitted topology of computable network for the application of the user (bus, ring, star, etc..).

A particular working of MASN is deduced from faults tolerance which has guided our study. The communications by wire with mobile robots is only considered in the following cases :
- when a failure occurs in the embedded energy supplying system; an external energy supply is necessary for the electronic devices and for recharging the batteries; the power line of the network then fulfills this role,
- in the implementation phase for fast teleloading of a software prototype or for debugging applications,
- when failure of the radio frequency system occurs.

2.3 The Generic Board

From this topologic description, we observe the specialized roles of the actors :
- the robotic actor, moving (MRA) or sedentary (SRA), which is an autonomous entity able to interact with its environment (passive objects and other actors),
- the basic hardware actor (BHA), whose presence is needed, in order to coordinate, if necessary, the development of an application,
- the communication manager hardware actor (CMHA), in charge of a bridge role between the MASN and the SASN, and
- the host computer (actually a PC), a bridge between the USN and the SASN.

This is represented by the following rules:
```
hardware_actor --> basic_hardware_actor I robotic_actor
        I communication_manager_hardware_actor
robotic_actor --> moving_robotic_actor I sedentary_robotic_actor
```

Each node in the network, or hardware actor, is controlled by a generic control board which permits occurrences for all kinds of actors. Indeed, the study of the control of hardware actors has shown that the architecture of the control board was similar and that only the input/output interface was specific [Darche 1990].

This is translated by the following rules :
```
hardware_package --> generic_board  [ pedagogical_robot ]
```

A pedagogical robot is characterized by :
```
pedagogical_robot        -->        moving_pedagogical_robot    I
sedentary_pedagogical_robot
moving_pedagogical_robot --> moving_commercial_robot
        I moving_prototype_robot
sedentary_pedagogical_robot --> sedentary_commercial_robot
        I sedentary_prototype_robot
```

Our choice is in favor of the following devices:
```
sedentary_commercial_robot --> robot_arm_Youpi
moving_prototype_robot --> MRA
```

The first actor we consider in the system is a basic hardware actor defined by:
    basic_hardware_actor --> software_package basic_hardware_actor_board
        communication_interface

If it is associated with a pedagogical robot, it becomes a robotic actor characterized
by the following rule:
    robotic_actor --> software_package generic_board pedagogical_robot

When instanciating this rule we obtain the following robotic actors:
    moving_robotic_actor --> software_package moving_robot_control_board
        moving_pedagogical_robot
    sedentary_robotic_actor --> software_package sedentary_robot_control_board
        sedentary_pedagogical_robot
communication_manager_hardware_actor -->
        software_package communication_manager_hardware_actor_board

The generic board is characterized by a micro-processor with a 16 or 32 bits
architecture, a RAM and ROM capacity from 512Kb to 4Mb by 512Kb banks and
by multiple input/output facilities (two bidirectional parallel interface units of
three 8 bits ports each, two full-duplex serial control units, and nine
programmable timers allowing, for example, an external clock management).

This genericity is defined by the following rule:
    generic_board --> CPU_16-32bits [RAM] [ROM] [I/O]
        [specific_interface]

It can be instanciated for the needs of the project as :
    minimal_control_board --> CPU_16-32bits
    basic_control_board --> minimal_control_board [ROM] [RAM]
    moving_robot_control_board -->
        basic_control_board [ROM] static_RAM I/O
        moving_robot_interface M.A.S.N._interface [S.A.S.N._interface]
    sedentary_robot_control_board -->
        basic_control_board I/O
        sedentary_robot_interface S.A.S.N._interface
    basic_hardware_actor_board --> basic_control_board S.A.S.N._interface
    communication_manager_hardware_actor_board -->
        basic_hardware_actor_board I/O S.A.S.N._interface
        M.A.S.N._interface M.A.S.N._interface

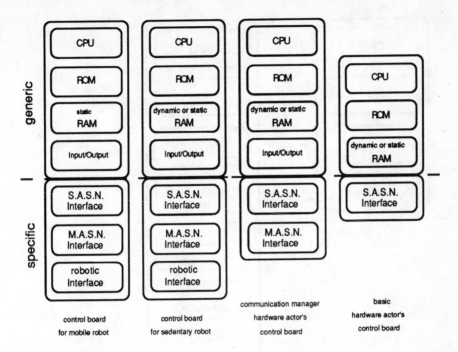

Fig. 2. Specialization of the Generic Board by Actor Type

The ROM is composed of an EPROM type memory for the prototype, but the standard models will be equiped of an EEPROM or FlashEPROM type memory allowing to keep the same physics principle (electrical) for the reading and the writing, therefore a greater development flexibility.

For a sedentary robot, the control board can either be equipped with static or dynamic RAM circuits. Only the performances and the cost are important for this choice (using of dynamic RAM gives an economy factor of 6, against a 30% access time loss). In other respects, the board of an MRA is imperiously equipped with static RAM to answer the constraints of consumption associated with energetic autonomy.

Two control architectures are proposed. The first one, a conventional one, still undergoing validation, is shown in Fig. 2.

The second one results from the fact that input/output, particularly the MRA ones, notably slow the CPU. The idea is therefore to free the CPU from input/output management. Input/output would be in charge of a specialized controller. This is shown by Fig. 3. This proposal enhances genericity and heterogeneity: control boards may be different, only I/O are unchanged. This flexibility of architecture thus permits fast hardware prototyping.

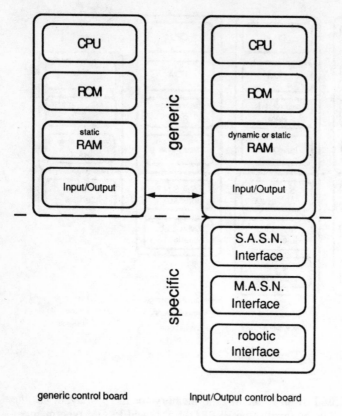

generic control board   Input/Output control board

Fig. 3. Second Architecture Proposal

2.4 Choice of Micro-Processor: The Transputer

Our application may be classified in the 'real-time' systems category, because it will have to react to such external events at the end of motor step interrupts, within a short time (about a millisecond).

The Transputers family [Inmos 89] from Inmos Company has been selected for the following features :
- the transputer is a component specially developed for distributed systems architectures,
- links are well suited for SASN giving a minimal cost for the interface (protocol being managed by the transputer itself),
- integrated fast memory (4Kb) increases the speed of execution by optimizing memory access,
- the RISC internal architecture of a transputer could lead to the expectation of high performances, mainly with an Occam compiler,
- integrated floating point unit (for the T800) enhances computing power and favours the local calculus of the coordinates of the robots (arm in space and the centre of the mobile in plane),

- CMOS technology implies a reasonable expenditure of energy with regard to the processor speed, and so allows it to be used as an embedded processor for MRA [Darche 90],
- initialization, teleloading and debugging can act with flexibility and efficiency with the possibility of booting with link or ROM,
- quasi-parallelism, thanks to the microprogrammed tasks scheduler, is well suited to the actors' behaviour. Furthermore, this integration permits fast processes management (queues, contexts management, priorities).

We end our description by writing the rules characterizing the components of the generic board :

CPU_16-32bits --> T414 | T425 | T800 | T805
RAM --> static_RAM | dynamic_RAM
ROM --> EPROM | EEPROM | Flash_EPROM
I/O --> discrete_I/O | I/O_processor
discrete_I/O --> parallel_interface_controller | serial_interface_controller | timer
I/O_processor --> CPU_8bits RAM ROM discrete_I/O
       |micro-controller [RAM] [ROM] [discrete_I/O]
CPU_8bits --> 680x | 80xx | Z80
micro-controller --> 68HC05 | 68HC11 | 80Cxxx
specific_interface --> robotic_interface | communication_interface
robotic_interface --> moving_robot_interface | sedentary_robot_interface
communication_interface --> M.A.S.N._interface | S.A.S.N._interface

Some new models of micro-processors, for example the Inmos T9000 [Inmos 91] or micro-processors from other families (Intel 80x86, Motorola 680x0, etc), may be considered to control the boards, with minimal hardware and software adjustments.

2.5 Economical Proposition

The technical choices made are implying that a high cost of realization, but for a pedagogical use in a school environment, the hardware and software actors' concepts will be implemented in an economical way. Each pedagogical robot will be equiped with a low cost command board (with a micro-controller) - in fact, the I/O_processor board equiped with a command interpreter. The robots will be regarded as peripherals. The notion of actors will be transfered at the level of a computer machine with a quasi-parallel system taken in the software actors and the pseudo-hardware actors (software actor plus a pedagogical peripheral), all that being of course transparent to the user.

3 Software Actors

3.1 Choice of Lisp

The system we propose is dedicated to self-apprenticeship. It therefore needs an interactive environment that offers activating, stopping and resuming processes, managing events, stepping users applications, and the managing of errors, eventually recovering after correction. An interpreter seems indispensable.

We have chosen Lisp for two aspects :
- as an assembly language, for high level implementers, extending towards objects [Masini et al., 1990] [Meyer 88] [AFCET 83 -> 88], actors [Agha 1986] [Bond- Gasser, 1988], agents [Ferber - Carle, 1990] models, expert systems and semantic nets,
- as a high level language for users (its syntax and semantics are easy, it is a functional language, possibly, by parts, redefinable and powerful because of recursivity).

3.2 Choice of Méta-Vlisp

We have chosen Méta-Vlisp, a "home made", available, open system issued from a grant with "National Education" [Saint-James, 1987]. Among its qualities is the fact that it is entirely written in Lisp. It automatically calculates a closure each time a lambda-expression is evaluated; this feature is very useful for object-oriented constructions. It generalizes the binding of arguments to parameters to tree-binding, which permits easy destructuration, for example, in analyzing messages.

We propose that the logical structure of a message is effectively a tree containing :
- a symbolic expression to be evaluated, and
- a list of senders,
- a list of receivers of the message and
- a list of receivers of the value resulting from evaluating the proposed expression.

Nowadays, the Méta-Vlisp system is evolving towards full C portability and has recently been upgraded with a partial evaluator.

For the past few years we have developed an interactive system using a moving "LOGO turtle-like" robot, controlled by a Lisp-based system and equipped, on board, with a reduced assembly language real time kernel. We are now substituting a complete Lisp interpreter for that reduced kernel. At present, we have made assembly versions and high level languages (C and shortly Ada) available for the following microprocessors : 80X86, 680X0, SPARC2 and Transputer.

3.3 Implementing Méta-Vlisp on Transputer

To implement Méta-Vlisp on the transputer, we first explored the capabilities of Occam, the basic language for the transputer : its value is that the produced code is very effective, but it has limited capabilities (very few data structures, no recursivity), and it is transputer-specific, thus cannot easily be used with external developments. Furthermore, its in-line assembly code mixing possibility, absolutely necessary for our work, was not sane enough to achieve this particular implementation!

The C language is not really standardized, but seems to be a good choice. At present time, however, code generated onto transputers seems to be relatively inefficient.

In this way we implemented the Méta-Vlisp system [Saint-James 90] using the C language ("C3L" compiler from the 3L company) onto transputer. Efficiency was rather poor, but we demonstrated that implementing this system in a high level language is possible, and we have got improved efficiency (13%) by using a new C compiler from Inmos "ICC", which is also characterized by being Ansi-C standard and a better performance for itself and the generated code. An implementation with "Helios-C" for a T-Node system is also complete.

After a great temptation to define all a System/Language, homogeneously, in Lisp, and after having verified those ideas [META 88], we distinguish now on the one hand, the real special features of Lisp and the need for dynamic functional redefinition, and on the other hand, purely 'system' aspects and environment management. It seems then that the main benefit of Lisp and particularly of Méta-Vlisp rests essentially with its "eval" function and its 300 or 500 predefined functions as functional functions, escaping, closure generation, trees management, and stack management which offers some optimizations by the iterative processing of recursively written code. Writing in Lisp, the Lisp environment: reader, printer, files processing, interactivity with debugging tools (trace, step, stack showing) are no longer justified in the context of a Lisp implementation on transputers, as for a net of actors. Indeed, such actors only know communication channels and traditional input/output links in a serial or parallel sense.

In addition, given the sending and receiving message frequencies in such a net, those actors must take advantage of a non-interpreted management of input/output flow.

3.4 ADA versus Lisp

An ADA development system from Alsys has recently become available for the transputer; we have explained our interest for this language [Le Roch - Nowak, 1990], therefore we are interested in exploiting its capabilities, mainly package specification and task managing, relayed by the hardware possibilities of the transputer. ADA, however, is not an interactive language; this reduces the advantage in using it for controlling robots in an apprenticeship context. An interpreted language is appropriate in this context.

A distributed application, executed onto the net of actors, consists of the cooperation of some actors with each of them executing an algorithm expressed in

Lisp. Therefore it is necessary to obtain primitives for communication and synchronization between actors. We wish to use ADA facilities [BARNES 88] for expressing these relationships between actors, so we are at present studying how to implement an actor by integrating the two systems: Lisp and ADA.

We propose putting two quasi-parallel tasks onto each hardware actor :
- the first one assumes all communications managing within the net, by buffering message exchanges,
- the second one contains messages recognition and execution of the relevant actions : we have to redefine, inside the generated Lisp system, the management of input and output primitives Read and Print and streams possibilities with links of transputers.

We are studying how to implement such an organization, by exploring two ideas :
- connecting a communication level into the Méta-Vlisp system, which now exists on transputer;
- implementing this Lisp system in ADA, in order to obtain a structured, homogeneous, easy-to-implement product, and to make available numerous services (libraries, task management, floating point calculus, use of packages for realizing the virtual basic machine of the Lisp interpreter...) offered by ADA system.

4 Two Applications as Examples

To illustrate with other applications such as the critical section or the rendez-vous described in [Darche 1991], an implementation and a prospective follow:

4.1 "Jeu Fou" Jig-saw Puzzle, 'Artus Games'

We shall illustrate our subject with the help of an application example, prompted by 'brain racking', concerning the realization of the assemblage of pieces of a jig-saw puzzle. Each piece of the puzzle is stuck onto a building block and robots can take these building blocks for different algorithms (producer-consumer protocols) and tasks management.

This puzzle consists of the assembly of 9 square cards in a square of 3 by 3 cards, according to associations. Each card supports on each of its sides a resource, anterior part (a_i, i : 1 to 4) or posterior one (p_i, i : 1 to 4) of a symbol. Four kinds of symbols are distinguished. A correct association results in building a symbol of a given type. A solution is found when twelve associations of the 3 by 3 cards square are right. More than one solution may exist.

The resources :   A solution :

```
    a2       a4       a4              a4       a2       a4
p3 1 a1  p3 2 a1  p1 3 a2        p3 6 a1  p1 7 a3  p3 2 a1
    p4       p2       p3              p4       p4       p2

    a2       a4       a4              a4       a4       a2
p3 4 a4  p3 5 a1  p3 6 a1        p1 3 a2  p2 9 a3  p3 4 a4
    p1       p2       p4              p3       p1       p1

    a2       a2       a4              a3       a1       a1
p1 7 a3  p1 8 a3  p2 9 a3        a2 8 p2  a2 1 p4  a4 5 p2
    p4       p2       p1              p1       p3       p3
```

An algorithm providing the solution, given the description of the building blocks, can be expressed in Lisp, either in the monolithic way, or with a communicating actors-oriented strategy. Fig. 4 gives the Lisp program for the last strategy tested in a mono-processor context.

This example has already been proposed for study and realization to engineers after just six months of training within computing science.

Fig. 4. Building Blocks Actors Strategy

```
; File "actcubes.mvl" with messages management
; still in embryo

; <message>: ( <selector> <addressees> <solution> )
; <selector>: "center" | "pole" | "angle"
; <addressees>: ( <actor> * )
; <actor>: {c1 to c9}
; <solution>: (<a-list-cubes> <a-list-grafts>)
; <a-list-cubes>: "incremental" solution to place cubes
; with indicator: 20=C; 0=N; 1=E; 2=S; 3=W;
;                 10=NE; 11=SE; 12=SW; 13=NW
; and value: {c1 to c9}
; <a-list-grafts>: resources points of grafts
;              of "incremental" solution
; with indicator: 0=N; 1=E; 2=S; 3=W
; and value: {-4 to -1, 1 to 4} to code 8 symbols
;         (4 anterior and 4 posterior ones)

(dl a_cube ((actor resources) (method addressees (cubes grafts)))
   (selectq method
     ("center" (send (add actor cubes) resources))
     ("pole" (foreach resources
                 (let ((possibilities (graft_pole actor grafts)))
                   (and possibilities
                     (send (add possibilities cubes)
                        (add possibilities grafts))))))
     ("angle" (foreach resources
                 (let ((possibilities (graft_angle actor grafts)))
                   (and possibilities
```

```
            (foreach possibilities
               (send (add possibilities cubes)
                  grafts))))))))

(dl send (cubes grafts)
   (let ((addressees (sub actor addressees)))
      (print (if (<= (length cubes) 4)
             (list "pole" addressees (list cubes grafts))
             (if (= (length cubes) 5)
                (list "angle"
                   addressees
                   (list cubes (reorganize grafts)))
                (if (<= (length cubes) 8)
                   (list "angle" addressees (list cubes grafts))
                   (list "show" '(server) cubes)))))))

(dl application ()   ; plus some other definitions of functions
   (dictionary)
   ; ...
   (a_cube c9 '("center" (c1 c2 c3 c4 c5 c6 c7 c8 c9) (nil nil)))
   ; ...)
```

; **Results of a simulated session**

? (application)

```
; center (c9 ((0 . 4) (1 . 3) (2 . -1) (3 . -2)))
(pole (c1 c2 c3 c4 c5 c6 c7 c8)
 (((20 . c9)) ((0 . 4) (1 . 3) (2 . -1) (3 . -2))))
; pole (c7 ((0 . 2) (1 . 3) (2 . -4) (3 . -1)))
(pole (c1 c2 c3 c4 c5 c6 c8)
 (((3 . c7) (20 . c9))
  ((3 (2 . 3) (0 . -1)) (0 . 4) (1 . 3) (2 . -1))))
(pole (c1 c2 c3 c4 c5 c6 c8)
 (((0 . c7) (20 . c9))
  ((0 (3 . -1) (1 . 3)) (1 . 3) (2 . -1) (3 . -2))))     ; ...
; pole (c3 ((0 . 4) (1 . 2) (2 . -3) (3 . -1)))
(angle (c2 c5 c6 c8)
 (((1 . c3) (2 . c1) (3 . c4) (0 . c7) (20 . c9))
  ((1 (0 . -1) (2 . 2)) (2 (1 . -4) (3 . 2))
   (3 (2 . 4) (0 . -3)) (0 (3 . -1) (1 . 3))))))
(angle (c2 c5 c6 c8)
 (((3 . c3) (2 . c1) (1 . c4) (0 . c7) (20 . c9))
  ((3 (2 . -3) (0 . 4)) (2 (1 . -4) (3 . 2))
   (1 (0 . 2) (2 . -1)) (0 (3 . -1) (1 . 3))))))
; angle (c2 ((0 . 4) (1 . 1) (2 . -2) (3 . -3)))
(angle (c5 c6 c8)
 (((11 . c2) (3 . c3) (2 . c1) (1 . c4) (0 . c7)
   (20 . c9))
  ((3 (2 . -3) (0 . 4)) (2 (1 . -4) (3 . 2))
   (1 (0 . 2) (2 . -1)) (0 (3 . -1) (1 . 3))))))
(angle (c5 c6 c8)
 (((10 . c2) (3 . c3) (2 . c1) (1 . c4) (0 . c7)
```

```
  (20 . c9))
  ((3 (2 . -3) (0 . 4)) (2 (1 . -4) (3 . 2))
   (1 (0 . 2) (2 . -1)) (0 (3 . -1) (1 . 3)))))        ; ...
; angle (c8 ((0 . 2) (1 . 3) (2 . -2) (3 . -1)))
(angle (c6)
 (((12 . c8) (10 . c5) (11 . c2)
   (3 . c3) (2 . c1) (1 . c4) (0 . c7) (20 . c9))
  ((3 (2 . -3) (0 . 4)) (2 (1 . -4) (3 . 2))
   (1 (0 . 2) (2 . -1)) (0 (3 . -1) (1 . 3)))))
(angle (c6)
 (((12 . c8) (11 . c5) (10 . c2)
   (3 . c3) (2 . c1) (1 . c4) (0 . c7) (20 . c9))
  ((3 (2 . -3) (0 . 4)) (2 (1 . -4) (3 . 2))
   (1 (0 . 2) (2 . -1)) (0 (3 . -1) (1 . 3)))))
; angle (c6 ((0 . 4) (1 . 1) (2 . -4) (3 . -3)))
(show (serveur)
 ((13 . c6) (12 . c8) (10 . c5) (11 . c2)
  (3 . c3) (2 . c1) (1 . c4) (0 . c7) (20 . c9)))
(show (serveur)
 ((13 . c6) (12 . c8) (11 . c5) (10 . c2)
  (3 . c3) (2 . c1) (1 . c4) (0 . c7) (20 . c9)))
```

4.2 Intuitive Sorts from Pupils

This application is taken directly from young pupils ideas. They were 9 to 11 years old in the final year of primary school [13].

The "instruction" was: 'You are ten in the classroom. I give you, each and secretly, a different number between 1 and 100. How do you imagine a way for all the pupils to be sorted by this number along a wall ?'.

We shall retain some rare but powerful strategies.

- Nadège: 'We ask the others their number and we take our place in order.'
- Olivier: 'If I had number 99, I'd wait for the preceding number to take my place.'
- Véronique: 'Each child has a number that he transmits to others and each classifies it in a list he has made.'
- Jérôme: 'We ask each for his number. If it is less than 50, we put him on one side, if it is greater than 50, we put him on the other side and we go again with 25 and 75.'
- Christophe: 'To sort in numerical order: we take the lowest and the highest among the numbers, we classify them, then we take again the lowest and the highest of the remaining numbers until there are no more numbers or only one number. If so we put him in the middle.'

Suppose we dispose of ten little mobile robots! (software actors could first be used to simulate, else).

A possible scene issued from Christophe's idea, may be organized as follows:

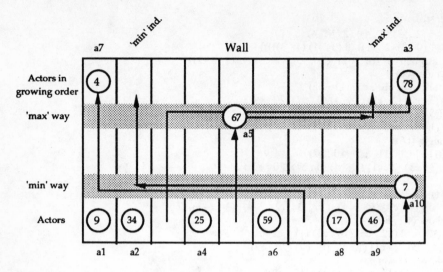

Fig. 5. Possible scene

One "actor initiator" is characterized by :
- its local knowledge :
    . the min and max boundaries of the numbers for the sort (resp. 1 and 100),
    . the pairs obtained by coupling :
        - the physical successive numbers of the actors (from 1 to 10) it is in fact a list L of actors which must still be sorted,
        - the logical random numbers, for the sort (inside circles on the figure),
    . the min and max boundaries and current index to place actors along the wall,
    . the information and structure of the messages (list L, current min and max physical and logical numbers, index where to place along the wall),
- its behaviour :
    . to initiate the primitive knowledge of the actors to be sorted,
    . to step the algorithm of the sort :
        - send to first actor of the list L, a solicitation to be less than actual min value and/or greater than actual max value, and
        - recognize the final condition (length of L is 0 or 1),
    . to broadcast to current min and max actors that they can go to resp. min and max places along the wall.

An "actor to be sorted" is characterized by :
- its local knowledge :
    . its physical number,
    . its logical number (for the sort),
    . the information and structure of the messages,

- its behaviour :
  . to learn its primitive knowledge,
  . to answer to solicitation of "being less than or greater than...", compare and send the same solicitation to first actor of L, if there still is one, otherwise send a message to actor initiator,
  . to go to min or max place along the wall.

This is a possible proposal among many. It can be specified with older pupils so as to avoid collisions during displacements between the mobiles by co-ordination messages.

5 Conclusion

The described architecture is presenting higher performances at the level of computational power, memory capacity, energetic autonomy, etc... The genericity and heterogeneity of our architecture allows the adaptation of different kinds of pedagogical robots and different kinds of control boards, so leading the evolving architecture towards the use of the T9000 and other processors (68X00 et 80X86 en particulier). Currently, a prototype of the control board, based on a T425, works for the moving robot, the control board with dynamic RAM for the sedentary robotic actor is still in development and the mobile prototype robot is under construction.

Many recent works have shown the matching of Lisp with domains as different as interrupts, processes or windows management, "objects oriented language" development, expert systems, etc.

The implementation of the Méta-Vlisp system by way of producing an assembly code generates an effectual version of the "eval" kernel but needs to dispose of some days, of the development system, and to know very precisely this context. We hope, from a C production of Méta-Vlisp, and more in ADA, for a real portability effect which will reduce to a matter of hours the delay in obtaining a new version.

On one hand, the compiler for C, C++, and ADA (our choice and interest) are often present on new processors, for many Operating Systems, and provide effectual libraries, in standardization phase and new features as objects or tasks.

On the other hand, with these languages, we may not be able to dispose of interactive facilities for an application being developed, unless we envelope it with an interpreter.

So, we think more in a synergy with interaction of the products Lisp and ADA for the future, which, as a beneficial side effect, must constrain more those tools to "open" on external relations. In this way, we have already developed in ADA a generic application for synchronizing tasks between two - producer and consumer - little robots. The associated Petri net for this work, designed with A.M.I. tools [Kordon et al. 1990] [Breant, 1990], allows us to obtain another corresponding ADA code. We are currently studying the required interface between these two applications.

Finally, we think that this project is a qualitative step in comparison with the traditional "LOGO turtle" micro-world. The notion of micro-world associated with the notion of software actors and hardware actors (robotic actor among others) are strong concepts which, implemented in a data processing system, are converting it into a powerful pedagogical tool.

Acknowledgements

Our numerous thanks are addressed to Marie-France Le Roch, Emmanuel Saint-James, Louis Audoire, Pierre Cointe, Philippe Gautron, Marc Cheminaud and Radu Greceanu.

References

AFCET (1988) 'AFCET-Informatique', reports of 'Objects Oriented Languages' group. Bigre + Globule (1983 to 1988)

Agha, A. (1986) Actors: A model of concurrent computation in distributed systems. The MIT Press, Cambridge, MA.

Barnes, J. (1988) Programmer en ADA. Inter Editions.

Bond, A., Gasser, L. (1988) Distributed Artificial Intelligence, Communication, Co-ordination. Morgan-Kaufmann, Lecture notes.

Bréant, F. (1990) TAPIOCA: Occam Rapid Prototyping From Petri Nets. Proceedings of the 5th Jerusalem Conference on Information Technology.

Darche, Ph. (1990) Conception et réalisation d'un environnement de contrôle d'acteurs robotiques. Université Pierre et Marie Curie, 3rd cycle DEA report.

Darche, Ph. (1991) Réseau d'acteurs robotiques pour l'apprentissage du parallélisme. Proceedings of the 3rd international Mexico conference on pedagogic robotics Mexico.

Darche, Ph., Nowak, G. (1992) ActNet: A Network of Actors based on Transputers for Apprenticeship of Parallelism. Proceedings of Transputers '92 (Advanced Research and Industrial Applications), Arc et Senans, May. IOS Press.

Ferber, J., Carle, P. (1990) Actors and Agents as Reflexive Concurrent Objects: a Mering IV perspective. Université Pierre et Marie Curie, IBP report n° 18/90.

Hewitt, C. (1977) Viewing Control Structures as Patterns of Passing Messages. Artificial Intelligence, 8 (3).

INMOS (1989) The Transputer data book. INMOS Corporation Ref. 72 TRN 203 01.

INMOS (1991) The T9000 Transputer products overview manual. INMOS Corporation Ref. 72 TRN 228 00.

Productique, J.D. (1986) Documentation du robot 'Youpi'. J.D. Productique.

Kordon, F., Estraillier, P., Card, R. (1990) Rapid ADA Prototyping: Principles and example of a complex application. Proceedings of the International Phoenix Conference on Computer and Communications.

Le Roch, M.F., G. Nowak, G. (1985) En amont de LOGO... Apprentissages au Cours Moyen: Communication, Coordination. Université Pierre et Marie Curie, experimentation report.

Le Roch, M.F., Nowak, G. (1990) Synchronisation de robots avec ADA. Université Pierre et Marie Curie 3rd cycle DESS project report.

Masini, G., Napoli, A., Colnet, D., Léonard, D., Tombre, K. (1990) Les langages à objets. Inter Editions.

Meyer, B. (1988) Object-Oriented Software Construction. Prentice Hall.

Papert, S. (1980) Mindstorms: children, computers and powerful ideas. Basic Books, New York.

Saint-James, E. (1987) De la Méta-Récursivité comme outil d'implémentation. Université Pierre et Marie Curie, Thèse d'Etat.

Saint-James, E. (1990) Un assembleur pour une Machine Endomorphique. Bull Company, research report.

Méta (1988) Méta-Vlisp. Manuel d'utilisation de l'interprète. Laboratoire LITP - Université Pierre et Marie Curie.

Glossary

ActNet	Actors' Network
A.M.I.	Atelier de Modélisation Interactif (Software Environment for Modeling Applications with Petri Nets)
B.H.A.	Basic Hardware Actor
B.M.F.S.K.	Binary Minimum Frequency Shift Keying
C.M.H.A.	Communication Manager Hardware Actor
F.M.	Frequency Modulation
M.A.S.N.	Moving Actors' Sub-Network
M.R.A.	Mobile Robotic Actor
S.A.S.N.	Sedentary Actors' Sub-Network
S.R.A.	Sedentary Robotic Actor
U.S.N.	Users' Sub-Network
V.H.F.	Very High Frequency

Participants and Contributors

Mr. David Anthony Argles
King Alfred's College
Sparkford Road
Winchester, Hants
SO22 4NR, United Kingdom

Mr. Jean-Claude Brès
Ecole active de Malagnou
39b, route de Malagnou
1208 Genève Switzerland

Mr. Jean-Claude Briers
Ecole primaire autonome
de la Communauté Française
2, rue de Void
4260 Ciplet, Belgium

Mr. Eduardo Calabrese
Dipartimento di Ingegneria
dell'Informazione
Viale delle Scienze
43100 Parma, Italy

Mr. Augusto Chioccariello
Istituto delle Tecnologie Didattiche
Via all'Opera Pia, 11
16145 Genova, Italy

Mr. Sean Close
St. Patricks College
Drumcondra
Dublin 9, Ireland

Mr. Philippe Darche
Université Pierre et Marie Curie
Paris 6, Labo. Masi Couloir 65-66
4, Place Jussieu
75252 Paris Cedex 05, France

Mrs. Brigitte Denis
Service de Technologie de
l'Education
University of Liège
Boulevard du Rectorat, 5 Bât. B32
Sart Tilman 4000 Liège 1, Belgium

Mr. Mike P. Doyle
British LOGO user group
37 Bright Street
Skipton, North Yorks BD23 1OO,
United Kingdom

Mr. Jorma Enkenberg
University of Joensuu
Research&Development Center ITE
P.O. Box 111
80101 Joensuu Finland

Mr. Reg Eyre
Wiltshire Education Authority
High Beech
Elkstone, Cheltenham
Glos GL53 9PA, United Kingdom

Mr. Gregory Gargarian
Institute of Technology
Massachusetts
20 Ames Street
02139 Cambridge, MA, USA

Mrs. Anne Gilbert
FNDP
Département Education et
Technologie
61, rue de Bruxelles
5000 Namur, Belgium

Mrs. Maria Lucia Giovannini
Dip. di Scienze dell'Educazione
Università Bologna
Via Zamboni, 34
40126 Bologna, Italy

Mrs. Montse Guitert
Universidad de Barcelona
Divisio de Ciènces de l'Educacio
Baldire Reixach, S/N Bloc D, pis 4rt
08028 Barcelona, Spain

Mr. Georges Gyftodimos
University of Athens
Department of Informatics
Panepistimiopolis, TYPA Buildings
15771 Ilisia, Greece

Mr. Carlo Krier
ISERP
2 BP
7701 Walferdange, Luxembourg

Mr. Jean-Baptiste La Palme
Université de Québec à Montréal,
Département Mathématique et
Informatique
Case postale 8888, Succursale "A"
Montréal, Québec, Canada H3C 3P8

Mr. Dieudonné Leclercq
Service de Technologie de
l'Education
University of Liège
Boulevard du Rectorat, 5 Bât. B32
Sart Tilman 4000 Liège 1, Belgium

Mr. Pascal Leroux
Université du Maine
Laboratoire d'Informatique
Faculté des Sciences
BP 535, Avenue Olivier Messiaen
72017 Le Mans Cedex, France

Mr. M. Louttit
Swallow-Systems
134, Cock Lane
High Wycombe, Bucks HP13 7EA,
United Kingdom

Mr. Joao Filipe Matos
Universidade de Lisboa
Departamento de Educaçao,
Faculdado de Ciencias
Av. 24 de Julho - 134 - 4
1300 Lisboa, Portugal

Mr. Maurice D. Meredith
40, Vectis Court
Talbot Close
Southampton SO1 7LY,
United Kingdom

Mr. Ole Moeller
LEGO DACTA
7190 Billund, Denmark

Mr. Pierre Nonnon
Université de Montréal
C.P. 6128, Succursale "A"
Montréal, Québec, H3C 3J7 Canada

Mr. Chris Robinson
Horndean C.E. Middle School
Five Heads Road
Horndean, Waterlooville
Hants P08 9NW, United Kingdom

Mrs. Marilyn Schaffer
Director of Educational Computing
University of Hartford
West Hartford, CT 06117; USA

Mr. Jacques Sougné
Service de Technologie de
l'Education
University of Liège
Boulevard du Rectorat, 5 Bât. B32
Sart Tilman 4000 Liège 1, Belgium

Mr. Martin Valcke
Open University of the Netherlands
Centre for Educational Technology
Valkenburgerweg 167
6419 Heerlen, The Netherlands

Mr. Martial Vivet
Faculté des Sciences
Université du Maine
BP 535
72017 Le Mans Cédex, France

Mr. Georges Gyftodimos
University of Athens
Department of Informatics
Panepistimiopolis, TYPA Buildings
15771 Ilisia, Greece

Mr. Carlo Krier
ISERP
2 BP
7701 Walferdange, Luxembourg

Mr. Jean-Baptiste La Palme
Université de Québec à Montréal,
Département Mathématique et
Informatique
Case postale 8888, Succursale "A"
Montréal, Québec, Canada H3C 3P8

Mr. Dieudonné Leclercq
Service de Technologie de
l'Education
University of Liège
Boulevard du Rectorat, 5 Bât. B32
Sart Tilman 4000 Liège 1, Belgium

Mr. Pascal Leroux
Université du Maine
Laboratoire d'Informatique
Faculté des Sciences
BP 535, Avenue Olivier Messiaen
72017 Le Mans Cedex, France

Mr. M. Louttit
Swallow-Systems
134, Cock Lane
High Wycombe, Bucks HP13 7EA,
United Kingdom

Mr. Joao Filipe Matos
Universidade de Lisboa
Departamento de Educaçao,
Faculdado de Ciencias
Av. 24 de Julho - 134 - 4
1300 Lisboa, Portugal

Mr. Maurice D. Meredith
40, Vectis Court
Talbot Close
Southampton, SO1 7LY,
United Kingdom

Mr. Ole Moeller
LEGO DACTA
7190 Billund, Denmark

Mr. Pierre Nonnon
Université de Montréal
C.P. 6128, Succursale "A"
Montréal, Québec, H3C 3J7 Canada

Mr. Chris Robinson
Horndan C.E. Middle School
Five Heads Road
Hordean Waterlooville
Hants, P08 9NW, United Kingdom

Mrs. Marilyn Schaffer
Director of Educational Computing
University of Hartford
West Hartford, CT 06117; USA

Mr. Jacques Sougné
Service de Technologie de
l'Education
University of Liège
Boulevard du Rectorat, 5 Bât. B32
Sart Tilman 4000 Liège 1, Belgium

Mr. Martin Valcke
Open University of the Netherlands
Centre for Educational Technology
Valkenburgerweg 167
6419 Heerlen, The Netherlands

Mr. Martial Vivet
Faculté des Sciences
Université du Maine
BP 535
72017 Le Mans Cédex, France

Printing: Druckhaus Beltz, Hemsbach
Binding: Buchbinderei Schäffer, Grünstadt

NATO ASI Series F

Including Special Programmes on Sensory Systems for Robotic Control (ROB) and on Advanced Educational Technology (AET)

NATO ASI Series F

Including Special Programmes on Sensory Systems for Robotic Control (ROB) and on Advanced Educational Technology (AET)

NATO ASI Series F

Including Special Programmes on Sensory Systems for Robotic Control (ROB) and on Advanced Educational Technology (AET)